Installing and Servicing Domestic Central Heating Wiring Systems and Controls

Installing and Servicing Domestic Central Heating Wiring Systems and Controls

Ray Ward

Newnes

AMSTERDAM BOSTON HEIDELBERG LONDON NEW YORK OXFORD
PARIS SAN DIEGO SAN FRANCISCO SINGAPORE SYDNEY TOKYO

Newnes
An imprint of Elsevier Science
Linacre House, Jordan Hill, Oxford OX2 8DP
200 Wheeler Road, Burlington, MA 01803

First published 1998
Reprinted 2000, 2001, 2002, 2003 (three times)

Copyright © 1999, Elsevier Science Ltd. All rights reserved.

No part of this publication may be reproduced in any material form (including photocopying or storing in any medium by electronic means and whether or not transiently or incidentally to some other use of this publication) without the written permission of the copyright holder except in accordance with the provisions of the Copyright, Designs and Patents Act 1988 or under the terms of a licence issued by the Copyright Licensing Agency Ltd, 90 Tottenham Court Road, London, England W1T 4LP. Applications for the copyright holder's written permission to reproduce any part of this publication should be addressed to the publisher.

Permissions may be sought directly from Elsevier's Science and Technology Rights Department in Oxford, UK: phone: (+44) (0) 1865 843830; fax: (+44) (0) 1865 853333; e-mail: permissions@elsevier.co.uk . You may also complete your request on-line via the Elsevier Science homepage (www.elsevier.com), by selecting 'Customer Support' and then 'Obtaining Permissions.

British Library Cataloguing in Publication Data
A catalogue record for this book is available from the British Library.

ISBN 0 7506 3994 6

Library of Congress Cataloguing in Publication Data
A catalogue record for this book is available from the Library of Congress.

For information on all Newnes publications
visit our website at www.newnespress.com

Typeset by David Gregson Associates, Beccles, Suffolk
Printed in Great Britain by Martins the Printers Ltd, Berwick upon Tweed

CONTENTS

	Acknowledgements	vi
1	Guide to use	1
2	Programmers and time switches	4
3	Programmers and time switches with inbuilt or external sensors or thermostats	62
4	Cylinder and pipe thermostats	74
5	Room, frost and low-limit thermostats	78
6	Motorized valves and actuators	91
7	Boilers – general	106
8	Boilers – gas	109
9	Boiler wiring – oil	156
10	Ancillary controls	163
11	Wiring system diagrams	182
12	Interchangeability guide for programmers and time switches	215
13	Manufacturers' trade names and directory	220
	Index	227

ACKNOWLEDGEMENTS

The author wishes to thank Meryl Brooks for her invaluable help in compiling the details required to produce this book. Also, the manufacturers who without exception have co-operated greatly and consented to the reproduction of their diagrams and illustrations. Thanks also to Richard Hawkes for his expertise in drawing the system wiring diagrams.

1

Guide to use

It is essential that this section is read and understood thoroughly prior to use of the book. All information supplied is believed to be correct, and as such no responsibility can be taken for errors or misuse of information.

General

All equipment is listed in numerical and alphabetical order within its own section, and where items are included elsewhere this is mentioned. An index at the back of the book gives additional information.

Manufacturers and trade names

Over the years manufacturers have merged or been taken over by other companies and every effort has been made to guide the reader to the correct location for information. However some items were manufactured under two names, e.g. Apollo boiler was produced by Myson and Thorn but Myson have since merged with Potterton, therefore a list is given below of where some difficulties may arise:

(a) ACL, Drayton, Grasslin, Motortrol, Switchmaster, Tower
(b) Danfoss, Randall
(c) Myson, Potterton, Thorn
(d) Satchwell, Sunvic

Reference to the manufacturers' and trade names directory in Chapter 13 will also help.

Programmers and time switches

These are listed in Chapter 2 in manufacturer order. The first detail is whether the item is electromechanical (driven by a motor) or electronic. Then the setting programme is indicated, i.e. 24 hour, 5/2 day (weekday/weekend), or 7 day. The terms 'basic' and/or 'full' are used if the item is a programmer. The term 'Basic' means that the Programmer does not have the facility for programming 'central heating only' and would be used, for example in a gravity hot water, pumped central heating system. The term 'Full' means that central heating can be selected without hot water such as is required for a normal fully pumped system. Programmers described as 'Basic/Full' have the facility for either option and details are given on how to alter the programmer as required. The maximum number of switching options (on/off), usually per day, are given as well as the current rating of the programmer switch in amps. The rating given will be for a resistive load and a rating for inductive loads may be given in brackets. Dimensions are also given in millimetres and this information can be extremely useful when having to replace an obsolete or unavailable model.

Programmers and time switches with inbuilt or external sensors or thermostats

Wiring and specification details are broadly similar to that given for the room thermostats and programmers.

Wiring system diagrams

Diagrams of all usual systems are included plus those of systems where special requirements may need to be met. All of the full system diagrams are based on the use of a junction box or wiring centre, although it is of course possible to connect wiring into a suitable programmer by following the wiring through. A full list of wiring diagrams included is given at the beginning of Chapter 11. **Important note:** for clarity all earth connections have been omitted but must be made where required.

Cylinder and pipe, room and frost thermostats

The room and frost thermostat details are listed separately to those of the cylinder and pipe thermostats, although the information given is similar. It can be assumed that all thermostats are suitable for 240V unless stated otherwise. The terminal identification is given as follows:

Common	The 'live in' terminal. In the case of a room thermostat, for example, this would be from the 'heating on' terminal of the programmer in most cases.
Demand	This contact will be 'made' to the common when the thermostat is calling or demanding heat.
Satisfied	This terminal will be 'made' to the common when the thermostat has reached the required temperature or is 'satisfied'.
Neutral	*Room thermostats* – should be wired where shown as this enables the heat anticipator to function and therefore make the thermostat more accurate and sensitive to alteration in temperature fluctuation. *Cylinder thermostats* – required for Potterton PTT1 and PTT2.

Also included are the available scale settings, the dimensions in millimetres, and the current rating of the thermostat contacts.

Motorized valves

Besides providing for wiring details of motorized valves and actuators as below, information is also included regarding port layout of 3-way valves, current rating of auxiliary switch, if fitted, and pipe sizes available.

Where the motorized valve or actuator is of a common type, for example 2-port spring return, diverter (3-port priority) or mid-position 3-port, then wiring may be given as *standard colour flex conductors* and these are as follows:

2-port 5-wire (or 4-wire without earth), spring return

Brown	Energize motor, usually to open valve
Blue	Neutral
Green/yellow	Earth (if fitted)
Orange	Live-in for auxiliary switch
Grey	Live-out from auxiliary switch when valve energized

Note: In the 2-wire auxiliary switch the orange and grey leads can be reversed. They may also have the colour coding black-black, white-white, or black-white, depending upon age and manufacturer.

2-port 6-wire (or 5-wire without earth), spring return, (excluding Sunvic SZ1302/2302)

As above with extra:

White	Live-out from auxiliary switch when valve de-energized. Orange and grey must be wired correctly as above. If this wire is spare then it must be made electrically safe.

3-port diverter – priority, spring return

Brown	Energize motor to open closed port (usually energized to open port to central heating)
Blue	Neutral
Green/yellow	Earth (if fitted)

3-port mid-position

Orange	From cylinder thermostat demand and to boiler and pump. Note that pump may need to be wired into boiler if boiler has pump over-run requirement.

White or brown	From programmer 'central heating on' via room thermostat if fitted.
Grey	From cylinder thermostat satisfied and also from 'hot water off' of programmer if possible. Without this second connection then 'central heating only' could not be selected if programmer is of the Full control type.
Blue	Neutral
Green/yellow	Earth (if fitted)

Boilers

The problem with boilers and associated information is deciding which ones should be included and how old. We have attempted to include all boilers that were still in production in 1983 and to date, therefore some boilers over 15 years old may be included, although it is felt that boilers beyond this time are unlikely to be incorporated into an updated system.

Besides wiring, the following information on boilers is included:

(a) Heat exchanger material
(b) Suitability for sealed systems
(c) Whether for fully pumped systems only
(d) Wall or floor mounted, or back boiler unit

The wiring of standard boilers is usually of two methods. Either a simple switched live, or, in the case of a boiler with pump over-run, a permanent live, switched live and pump live. Some back boiler units may require a permanent live to enable the bulbs on the fire front to work when the boiler is off. The wiring of combination boilers is usually via a voltage free switch of a time clock.

Ancillary equipment

Brief details of domestic compensator systems, boiler energy controls and similar are given and these are listed at the beginning of the section.

2

Programmers and time switches

ACL FP

```
1   2   3   4   5   6   7   8   9   10  11
O   O   O   O   O   O   O   O   O   O   O
L   N               HW      HW      CH  CH
MAINS               ON      OFF     ON  OFF
```

(a) Fully pumped 2 × 2 port motor open/close valves links 4–9, 5–7.
(b) Fully pumped 2 × 2 port spring return valves, 1 × 3-way mid-position valve, Satchell Duoflow Switchmaster Midi and Drayton Flowshare link 1–4–9 and 5–7
(c) Tower or ACL Biflo mid-position valve link 1–4–9
(d) Terminal 3 is a spare terminal

Electromechanical 24 hour Full programmer

Clock module available as a spare

On/off × 4
H106 × W113 × D65
Switch rating 6A

ACL MP

```
1   2   3   4   5   6   7   8   9   10  11
O   O   O   O   O   O   O   O   O   O   O
L   N               HW              CH
MAINS               ON              ON
```

(a) Link 1–4 and 6–11 for all systems except (b)
(b) Tower or ACL Biflo mid-position valve link 1–4
(c) Terminals 3, 7 and 8 are spare terminals

Electromechanical 24 hour Basic programmer

Clock module available as a spare

On/off × 4
H106 × W113 × D65
Switch rating 6A

ACL TC

```
1   2   3   4   5   6   7   8   9   10  11
O   O   O   O   O   O   O   O   O   O   O
L   N       see         ON
MAINS       note
```

(a) Link L–4–6 for 240V control
(b) Link 4–6 for voltage free switching – input to terminal 4
(c) terminals, 2, 8, 9, 10 and 11 are spare terminals

Electromechanical 24 hour time switch

Clock module available as a spare

On/off × 4
H160 × W113 × D65
Switch rating 10A

ACL TC/7

As TC with 7-day clock fitted

Programmers and time switches

ACL LP 111

N	L	1	2	3	4
○	○	○	○	○	○
MAINS		COM	OFF	ON	SPARE

Voltage free switching unless L–1 linked

Electronic 24 hour time switch

On/off × 2
H93 × W148 × D31
Switch rating 2A (1A)

ACL LP 112

N	L	1	2	3	4
○	○	○	○	○	○
MAINS		HW OFF	CH OFF	HW ON	CH ON

Electronic 24 hour Basic/Full programmer

On/off × 2
H93 × W148 × D31
Switch rating 2A (1A)
Move slider at rear of programmer to G for Basic control or P for Full control

ACL LP 241

N	L	1	2	3	4
○	○	○	○	○	○
MAINS		HW OFF	CH OFF	HW ON	CH ON

Facility for setting hot water and heating at different times in Full mode

Electronic 24 hour Basic/Full programmer

On/off × 2
H93 × W148 × D31
Switch rating 2A (1A)
Move slider at rear of programmer to G for Basic control or P for Full control

ACL LP 522

N	L	1	2	3	4
○	○	○	○	○	○
MAINS		HW OFF	HW OFF	CH ON	CH ON

Facility for setting hot water and heating at different times in Full mode

Electronic 5/2 day Basic/Full programmer

On/off × 2
H93 × W148 × D31
Switch rating 2A (1A)
Move slider at rear of programmer to G for Basic control or P for Full control

ACL LP 711

N	L	1	2	3	4
○	○	○	○	○	○
MAINS		COM	OFF	ON	SPARE

Voltage free switching unless L–1 linked

Electronic 7 day time switch

On/off × 2
H93 × W148 × D31
Switch rating 2A (1A)

ACL LP 722

N	L	1	2	3	4
○	○	○	○	○	○
MAINS		HW OFF	CH OFF	HW ON	CH ON

Facility for setting hot water and heating at different time from each other every day in Full mode

Electronic 7 day Basic/Full programmer

On/off × 2
H93 × W148 × D31
Switch rating 2A (1A)
Move slider at rear of programmer to G for Basic control or P for Full control

ACL LS 111

N	L	1	2	3	4
○	○	○	○	○	○
MAINS		COM	OFF	ON	SPARE

Voltage free switching unless L–1 linked

Electronic 24 hour time switch

On/off × 2
H81 × W165 × D44
Switch rating 2A(1A)

ACL LS 112

N	L	1	2	3	4
○	○	○	○	○	○
MAINS		COM	OFF	HW ON	CH ON

Voltage free switching unless L–1 linked

Electronic 24 hour Basic programmer

On/off × 2
H81 × W165 × D46
Switch rating 2A (1A)

ACL LS 241

N	L	1	2	3	4
○	○	○	○	○	○
MAINS		HW OFF	CH OFF	HW ON	CH ON

Electronic 24 hour Basic/Full programmer

On/off × 2
H87 × W170 × D47
Switch rating 2A (1A)
Turn screw at rear of programmer to G for Basic control or P for Full control

ACL LS 522

N	L	1	2	3	4
○	○	○	○	○	○
MAINS		HW OFF	CH OFF	HW ON	CH ON

Facility for 5/2 day setting

Electronic 24 hour Basic/Full programmer

On/off × 2
H87 × W170 × D47
Switch rating 2A (1A)
Turn screw at rear of programmer to G for Basic control or P for Full control

ACL LS 711

N	L	1	2	3	4
○	○	○	○	○	○
MAINS		COM	OFF	ON	SPARE

Voltage free switching unless L–1 linked

Electronic 7 day time switch

On/off × 2
H87 × W170 × D47
Switch rating 2A (1A)

ACL LS 722

N	L	1	2	3	4
○	○	○	○	○	○
MAINS		HW OFF	CH OFF	HW ON	CH ON

Facility for setting hot water and heating at different time from each other every day in Full mode

Electronic 7 day Basic/Full programmer

On/off × 2
H87 × W170 × D47
Switch rating 2A (1A)
Turn screw at rear of programmer to G for Basic control or P for Full control

Programmers and time switches

ACL 2000

As Tower T 2000

Barlo EPR1

As ACL LS 522

Crossling Controller

As Landis & Gyr RWB2

Danfoss 3001

As Horstmann 425 Coronet

Danfoss 3002

As Horstmann 425 Diadem

Danfoss CP 15

E	N	L	1	2	3	4	5	6
○	○	○	○	○	○	○	○	○
MAINS			HW OFF	CH OFF	HW ON	CH ON	SPARE	

Electronic 24 hour or 5/2 day Basic/Full programmer

On/off × 3
H88 × W135 × D38
Switch rating 3A (1A)

Danfoss CP 75

E	N	L	1	2	3	4	5	6
○	○	○	○	○	○	○	○	○
MAINS			HW OFF	CH OFF	HW ON	CH ON	SPARE	

Electronic 7 day or 5/2 day Basic/Full programmer

On/off × 3
H88 × W135 × D38
Switch rating 3A (1A)

Danfoss FP 15

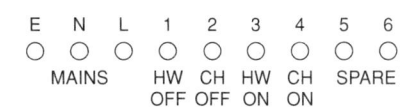

E	N	L	1	2	3	4	5	6
○	○	○	○	○	○	○	○	○
MAINS			HW OFF	CH OFF	HW ON	CH ON	SPARE	

Electronic 24 hour or 5/2 day Basic/Full programmer

On/off × 3
H88 × W135 × D38
Switch rating 3A (1A)

Facility for setting hot water and central heating at different times to each other

Danfoss FP 75

E	N	L	1	2	3	4	5	6
○	○	○	○	○	○	○	○	○
MAINS			HW OFF	CH OFF	HW ON	CH ON	SPARE	

Electronic 24 hour or 5/2 day Basic/Full programmer

On/off × 3
H88 × W135 × D38
Switch rating 3A (1A)

Facility for setting hot water and central heating at different times to each other

Danfoss MP 15

```
E   N   L   1   2   3   4   5   6
O   O   O   O   O   O   O   O   O
    MAINS       HW  CH  HW  CH  SPARE
                OFF OFF ON  ON
```

Danfoss MP 75

```
E   N   L   1   2   3   4   5   6
O   O   O   O   O   O   O   O   O
    MAINS       HW  CH  HW  CH  SPARE
                OFF OFF ON  ON
```

Danfoss TS 15

```
    N   L   1   2   3   4
    O   O   O   O   O   O
    N   L   COM OFF SPARE ON
    MAINS
```

Voltage free switching unless L–1 linked

Danfoss TS 75

```
    N   L   1   2   3   4
    O   O   O   O   O   O
    N   L   COM OFF SPARE ON
    MAINS
```

Voltage free switching unless L–1 linked

Danfoss Randall SET 1E

```
E   N   L       1   2   3       4   5   6
O   O   O       O   O   O       O   O   O
    MAINS           SPARE           OFF COM ON
```

Voltage free switching unless L–5 linked

Danfoss Randall SET 2E

```
E   N   L   1   2   3   4   5   6
O   O   O   O   O   O   O   O   O
    MAINS   HW  COM HW  CH  COM CH
            ON      OFF ON      OFF
```

Voltage free switching unless L–2–5 linked

Electronic 24 hour or 5/2 day Basic programmer

On/off × 3
H88 × W135 × D38
Switch rating 3A (1A)

Electronic 7 day or 5/2 day Basic programmer

On/off × 3
H88 × W135 × D38
Switch rating 3A (1A)

Electronic 24 hour or 5/2 day time switch

On/off × 3
H88 × W135 × D38
Switch rating 3A (1A)

Electronic 7 day or 5/2 day time switch

On/off × 3
H88 × W135 × D38
Switch rating 3A (1A)

Electronic 24 hour time switch

On/off × 2
H98 × W158 × D36
Switch rating 3 (1A)

Electronic 24 hour Basic programmer

On/off × 2
H98 × W158 × D36
Switch rating 3 (1A)

Programmers and time switches

Danfoss Randall SET 3E

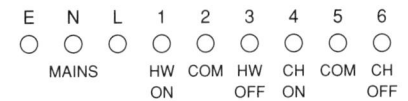

E	N	L	1	2	3	4	5	6
○	○	○	○	○	○	○	○	○
	MAINS			HW ON	COM	HW OFF	CH ON	COM	CH OFF

Voltage free switching unless L–2–5 linked

Electronic 24 hour Basic/Full programmer

On/off × 2
H98 × W158 × D36
Switch rating 3 (1A)

Danfoss Randall SET 3M

E	N	L	1	2	3	4	5	6
○	○	○	○	○	○	○	○	○
	MAINS 240V			HW ON	COM	HW OFF	CH ON	COM	CH OFF

Voltage free switching unless L–2–5 linked

Electromechanical 24 hour Basic/Full programmer

On/off × 2
H98 × W158 × D63
Switch rating 3A
Fit link supplied for Basic control

Danfoss Randall FP 975

E	N	L	1	2	3	4	5	6
○	○	○	○	○	○	○	○	○
	MAINS			HW OFF	COM	HW ON	CH OFF	COM	CH ON

Voltage free switching unless L–2–5 linked
With facility for 5/2 day setting

Electronic 7 day Basic/Full programmer

On/off × 3
H99 × W150 × D42
Switch rating 3A (1A)
Move slider at rear for Basic control

Danfoss Randall TS 975

E	N	L	1	2	3	4	5	6
○	○	○	○	○	○	○	○	○
	MAINS			SPARE		OFF	COM	ON

Voltage free switching unless L–5 linked
With facility for 5/2 day setting

Electronic 7 day time switch

On/off × 3
H99 × W150 × D42
Switch rating 3A (1A)

Drayton Tempus 1

E	N	L	1	2	3	4
○	○	○	○	○	○	○
	MAINS		COM	ON	OFF	SPARE

Voltage free switching unless L–1 linked

Electronic 24 hour time switch

On/off × 2
H84 × W140 × D46
Switch rating 3A (1A)

Drayton Tempus 2

E	N	L	1	2	3	4
○	○	○	○	○	○	○
	MAINS		COM	ON	OFF	SPARE

Voltage free switching unless L–1 linked
With facility for 5/2 day setting

Electronic 24 hour time switch

On/off × 2
H84 × W140 × D46
Switch rating 3A (1A)

Drayton Tempus 3

E	N	L	1	2	3	4
O	O	O	O	O	O	O
	MAINS		HW OFF	CH OFF	HW ON	CH ON

Electronic 24 hour Basic/Full programmer

On/off × 2
H84 × W140 × D46
Switch rating 3A (1A)
For Basic control remove plug from rear of programmer

Drayton Tempus 4

E	N	L	1	2	3	4
O	O	O	O	O	O	O
	MAINS		HW OFF	CH OFF	HW ON	CH ON

Facility for 5/2 day setting

Electronic 24 hour Basic/Full programmer

On/off × 2
H84 × W140 × D46
Switch rating 3A (1A)
For Basic control remove plug from rear of programmer

Drayton Tempus 7

E	N	L	1	2	3	4
O	O	O	O	O	O	O
	MAINS		HW OFF	CH OFF	HW ON	CH ON

Facility for setting hot water and heating at different times daily in Full mode

Electronic 7 day Basic/Full programmer

On/off × 2
H84 × W140 × D46
Switch rating 3A (1A)
For Basic control remove plug from rear of programmer

Eberle 606

E	1	2	3	4	5	6	7
O	O	O	O	O	O	O	O
E	L	N	HW ON	SPARE	CH ON	SPARE	N
	MAINS						

Terminals 2–7 are internally linked

Electromechanical 24 hour Basic programmer

On/off × 2

Eberle 607

E	1	2	3	4	5	6	7
O	O	O	O	O	O	O	O
E	L	N	HW ON	SPARE	CH ON	SPARE	N
	MAINS						

Terminals 2–7 are internally linked

Electromechanical 24 hour time switch with pump switch

On/off × 2

Eberle 608

For diagram see Figure 2.13, page 58

Electromechanical 24 hour priority programmer

On/off × 2

Eberle 609

See Eberle 633

Eberle 610 and 610/15

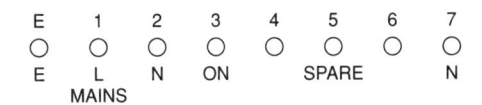

E	1	2	3	4	5	6	7
○	○	○	○	○	○	○	○
E	L	N	ON		SPARE		N
	MAINS						

Terminals 2–7 are internally linked

Electromechanical 24 hour time switch

On/off × 2

Eberle 633 (supercedes 609)

E	1	2	3	4	5	6	7
○	○	○	○	○	○	○	○
E	L	N	HW	HW	CH	CH	L
	MAINS		ON	OFF	ON	OFF	

Link L–7 unless used in conjunction with Honeywell V4073 6-wire mid-position valve (with external relay)

Electromechanical 24 hour Full programmer

On/off × 2
H193 × W105 × D72

Flash 31031 (FP 124)

N	N	L	1
○	○	○	○
N	MAINS		ON

Electromechanical 24 hour time switch

On/off × 36
H84 × W167 × D44
Switch rating 6A

Flash 31032 (FP 224)

N	N	L	1	2
○	○	○	○	○
	MAINS		HW	CH
			ON	ON

Electromechanical 24 hour Basic programmer

On/off × 36
H84 × W167 × D44
Switch rating 6A

Flash 31033 (FP 324)

N	N	L	1	2
○	○	○	○	○
N	MAINS		ON	ON

Electromechanical 24 hour Full programmer

On/off × 36
H84 × W167 × D44
Switch rating 6A

Flash 31731 (FP 17)

7 day version of 31031 with up to 6 on/offs per day

Flash 31731 (FP 27)

7 day version of 31032 with up to 6 on/offs per day

Flash 31733 (FP 37)

7 day version of 31033 with up to 6 on/offs per day

Glow-Worm M2525

```
 1    2    3    4    5
 O    O    O    O    O
 L    N    E    CH   HW
     MAINS      ON   ON
```

Electromechanical 24 hour Basic programmer

On/off × 2
H118 × W209 × D55

Glow-Worm Mastermind

As Landis & Gyr RWB2

Grasslin QE 1

See Tower QE1

Grasslin QE 2

See Tower QE2

Harp HGC1

```
 1         2      3    4    5    6    7    8    E    N    L
 O         O      O    O    O    O    O    O    O    O    O
GAS VALVE  HW    COM   HW   CH   COM  CH   E    N    L
FOR COST   ON         OFF   ON        OFF       MAINS
MONITORING
```

Voltage free switching unless L–4–7 linked

Electronic check cost programmer

On/off × 2
H134 × W205 × D48
Switch rating 5A (2A)

Hawk HTC1

See Switchmaster 980

Heatwave GP1/A

Electromechanical 24 hour Basic programmer

Heatwave GP2/A

Electromechanical 24 hour Basic programmer

Heatwave A1F/1A

Electromechanical 24 hour Full programmer

Honeywell ST499A

```
  8        6    5         3    N    L
  O        O    O         O    O    O
 COM       HW  COM        CH   N    L
           ON             ON       MAINS
```

Voltage free switching unless L–5–8 linked

Electronic 24 hour Full programmer

With off/timed/continuous options

On/off × 2
H100 × W100 × D38
Switch rating 2A (2A)

Honeywell ST699B

As ST699C, with off/once/twice/continuous options

Programmers and time switches

Honeywell ST699C

8	7	6	5	4	3	N	L
○	○	○	○	○	○	○	○
COM	HW OFF	HW ON	COM	CH OFF	CH ON	N MAINS	L

Voltage free switching unless L–5–8 linked

Electronic 24 hour Full programmer

With off/timed/continuous options

On/off × 2
H100 × W100 × D38
Switch rating 2A (1A)

Honeywell ST799

7 day version of ST699B

Honeywell ST6100A

N	L	1	2	3	4
○	○	○	○	○	○
N MAINS	L	COM	OFF	SPARE	ON

Voltage free switching unless L–1 linked

Electronic 24 hour time switch

On/off × 3
H95 × W145 × D52
Switch rating 3A (3A)

Honeywell ST6100C

N	L	1	2	3	4
○	○	○	○	○	○
N MAINS	L	COM	OFF	SPARE	ON

Voltage free switching unless L–1 linked

Electronic 7 day time switch

On/off × 3
H95 × W145 × D52
Switch rating 3A (3A)

Honeywell ST6200A

N	L	1	2	3	4
○	○	○	○	○	○
N MAINS	L	HW OFF	CH OFF	HW ON	ON

Electronic 24 hour Basic programmer

On/off × 2
H95 × W145 × D52
Switch rating 3A (3A)

Honeywell ST6300A

N	L	1	2	3	4
○	○	○	○	○	○
N MAINS	L	HW OFF	CH OFF	HW ON	CH ON

Electronic 24 hour Full programmer

On/off × 2
H95 × W145 × D52
Switch rating 3A (3A)

Honeywell ST6400C

N	L	1	2	3	4
○	○	○	○	○	○
N MAINS	L	HW OFF	CH OFF	HW ON	CH ON

Facility for setting hot water and heating at different times from each other

Electronic 7 day Full programmer

On/off × 3
H95 × W145 × D52
Switch rating 3A (3A)

Installing and Servicing Domestic Central Heating Wiring Systems and Controls

Honeywell ST6450 — Electronic 5/2 day Full programmer

N	L	1	2	3	4
N	L	HW OFF	CH OFF	HW ON	CH ON
MAINS					

On/off × 3
H95 × W145 × D52
Switch rating 3A (3A)

Facility for setting hot water and heating at different times

Honeywell ST7000A — Electronic 24 hour Basic programmer

CH ON	○	4
HW ON	○	3
HW OFF	○	2
LIVE	○	L

On/off × 2
H95 × W122 × D27
Switch rating 2A (2A)

The unit is battery powered and so no neutral is required

Honeywell ST7000B — Electronic 24 hour time switch

ON	○	3
OFF	○	2
LIVE	○	L

On/off × 2
H95 × W122 × D27
Switch rating 2A (2A)

The unit is battery powered and so no neutral is required

Honeywell ST7100 — Electronic 24 hour Full programmer

8	7	6	5	4	3
HW ON	HW OFF	COM	CH ON	CH OFF	COM
				N	L
○	○	○	○	○	○
SPARE				MAINS	

Facility for setting hot water and heating at different times to each other during 5/2 day

On/off × 3
H95 × W150 × D49
Switch rating 2A (2A)

Voltage free switching unless L–3–6 linked Terminals are provided for earth and neutral connections

Horstmann 401/K/1 — Electromechanical 24 hour time switch

401/K/1 Centaur

Time switch with voltage-free contacts only

N	L	E	N	N	L	1	2	3
N	L	E	N	N	COM	ON		
MAINS								

Any wires in 2 and 3 should be joined together in a spare terminal.
Do not link live to switch common in new clock unless done in old clock

Programmers and time switches

Horstmann 404/K/1   Electromechanical 24 hour time switch

404/K/1 Centaur	E	E	N	L	1	2	3	4	5	6	7	8	9	10	11	12
		MAINS														
Time switch with voltage-free contacts only	E	E	N	L	N			ON			COM		COM			
		MAINS														

Any wires in 3, 5, 6, 11 and 12 should be joined together in a spare terminal.
Any wires in 2 and 10 should be joined together in a spare terminal.
Do not link live to switch common in new clock unless done in old clock.

Horstmann 404/Q/4   Electromechanical 24 hour Basic programmer

404/Q/4 Centaur	E	E	N	L	1	2	3	4	5	6	7	8	9	10	11	12
		MAINS														
Basic programmer	E	E	N	L	N			CH		HW		N				
		MAINS														

Any wires in 3, 5, 11 and 12 should be joined together in a spare terminal.
Any wires in 2 and 10 should be joined together in a spare terminal.

Horstmann 423 Amber   Electromechanical 24 hour Full programmer

Designed for use on fully pumped system using change over thermostats and motor open/close motorized valves without end switches.

On/off × 2
H177 × W85 × D57
Switch rating 6A

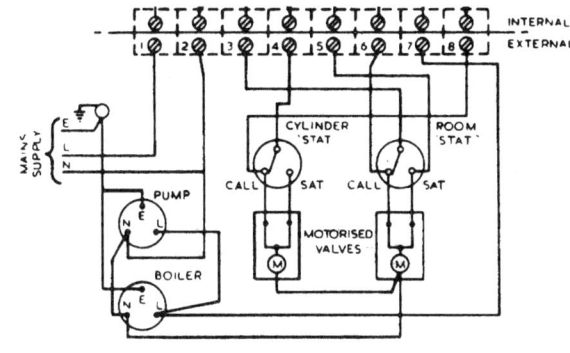

Figure 2.1

Horstmann 423 Amethyst 7+10   Electromechanical 24 hour Full programmer

1	2	3	4	5	6	7	8
L	N	N	HW OFF	HW ON	CH OFF	CH ON	SPARE
MAINS							

Amethyst 7 has off/constant/auto control
Amethyst 10 has off/constant/twice/all day control

On/off × 2
H177 × W85 × D57
Switch rating 6A

Horstmann 423 Coral

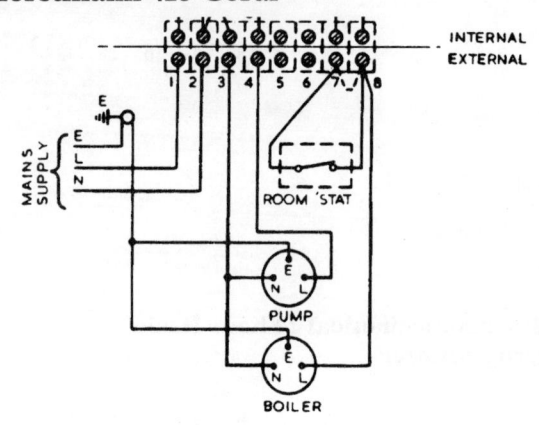

[Wiring diagram showing mains supply, terminals 1-8 internal/external, room stat, pump, and boiler connections]

Figure 2.2

Electromechanical 24 hour Basic programmer

On/off × 2
H177 × W85 × D57
Switch rating 6A

If a room thermostat is to be fitted remove link 7–8

Horstmann 423 Diamond

```
  5      N      1      2
  O      O      O      O
SPARE    N      L     HW
                      ON
  6      L      3      4
  O      O      O      O
SPARE    L      L     CH
                      ON
```

Electromechanical 24 hour Basic programmer

On/off × 2
H105 × W83 × D57
Switch rating 6A (2A)

Terminals L–1–3 have a bridging link which can be removed to provide separate switch and motor terminal connections.
Terminals 5 and 6 are provided for linking and have no internal connections to the time control.

Horstmann 423 Emerald

```
  N      1      2      5
  O      O      O      O
  N    SPARE  SPARE  SPARE

  L      3      4      6
  O      O      O      O
  L     COM    ON    SPARE
```

Terminals L–3 are linked internally but this can be removed for voltage-free switching

Electromechanical 24 hour time switch

On/off × 2
H105 × W83 × D57
Switch rating 6A (2A)

Horstmann 423 Leucite 10

```
  1    2    3    4    5    6    7    8
  O    O    O    O    O    O    O    O
  L    N   HW   HW  COM   CH  COM   CH
 MAINS     ON  OFF       ON        OFF
```

Link 5–7

Electromechanical 24 hour Full programmer

On/off × 2
H177 × W85 × D57
Switch rating 6A (2A)

Horstmann 423 Pearl 6

```
N     1      2      5
O     O      O      O
N   SPARE  SPARE  SPARE

L     3      4      6
O     O      O      O
L    COM     ON   SPARE
```

Terminals L–3 are linked internally but this can be removed for voltage-free switching

Electromechanical 24 hour time switch

On/off × 2
H105 × W88 × D57
Switch rating 6A (2A)

Horstmann 423 Pearl 16

As Pearl 6, but 16A (3A) switch rating

Horstmann 423 Pearl Auto 6 and 16

As Pearl, with off/constant/auto control
See also SMC programmers

Horstmann 423 Ruby

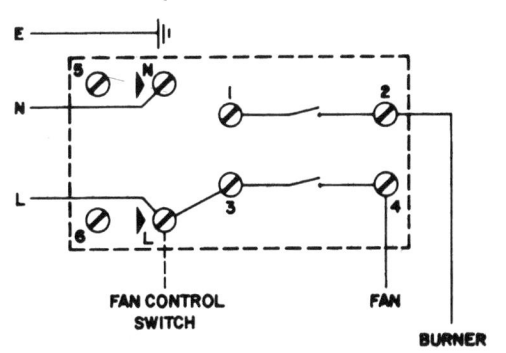

Electromechanical 24 hour time switch

Specifically designed for warm air units

On/off × 2
H105 × W88 × D57
Switch rating 6A (2A)

Figure 2.3

Horstmann 423 Sapphire

For diagram see Figure 2.14, page 58

Electromechanical 24 hour priority programmer

On/off × 2
H177 × W85 × D57
Switch rating 6A

Horstmann 423 Topaz

Specifically designed for night set-back thermostat

Electromechanical 24 hour time switch

On/off × 2
H105 × W88 × D57
Switch rating 6A (2A)

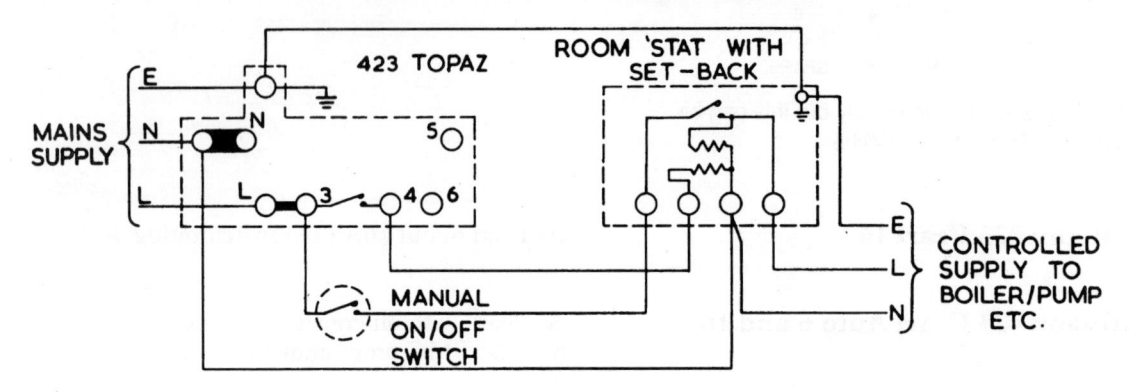

Figure 2.4

Horstmann 424 Amber

For diagram see Figure 2.1

Electromechanical 24 hour Full programmer

Designed for use on fully pumped system, using change over thermostats and motorized valves without end switches

On/off × 2
H177 × W85 × D57
Switch rating 6A

Horstmann 424 Amethyst 7

1	2	3	4	5	6	7	8
L	N	N	HW OFF	HW ON	CH OFF	CH ON	SPARE
MAINS							

Electromechanical 24 hour Full programmer

On/off × 2
H177 × W85 × D57
Switch rating 6A

Horstmann 424 Coral

For diagram see Figure 2.2

Electromechanical 24 hour Basic programmer

On/off × 2
H177 × W85 × D57
Switch rating 6A

Horstmann 424 Diamond

```
              L1        2
              O         O
      MAINS   N         HW
                        ON
        N     3    4    5
        O     O    O    O
      MAINS   N    L    CH    SPARE
                        ON
```

Terminals L–3 are linked internally

Horstmann 424 Emerald

```
              L1        2
              O         O
      MAINS   L         SPARE

        N     3    4    5
        O     O    O    O
      MAINS   N    COM  ON    SPARE
```

Terminals L1–3 are linked internally but this can be removed for voltage-free switching

Horstmann 424 Gem

```
  1   2   3    4     5     6    7    8    9     10
  O   O   O    O     O     O    O    O    O     O
  L   N   N    HW    COM   HW   CH   COM  CH    L
MAINS          ON          OFF  ON        OFF
```

Link 5–8–10

Horstmann 424 Leucite

```
  1   2    3     4     5    6    7     8
  O   O    O     O     O    O    O     O
  L   N    HW    HW    COM  CH   COM   CH
MAINS      ON    OFF        ON         OFF
```

Link 5–7

Horstmann 424 Pearl

```
              L1        2
              O         O
      MAINS   L         SPARE

        N     3    4    5
        O     O    O    O
      MAINS   N    COM  ON    SPARE
```

Terminals L1–3 are linked internally but this can be removed for voltage-free switching

Electromechanical 24 hour Basic programmer

On/off × 2
H130 × W87 × D57
Switch rating 6A (2A)

Electromechanical 24 hour time switch

On/off × 2
H130 × W87 × D57
Switch rating 16A (3A)

See also SMC programmers

Electromechanical 24 hour Full programmer

On/off × 2
H177 × W86 × D57
Switch rating 6A

Electromechanical 24 hour Full programmer

On/off × 2
H177 × W86 × D57
Switch rating 6A

Electromechanical 24 hour time switch

On/off × 2
H130 × W87 × D57
Switch rating 16A (3A)

Horstmann 424 Pearl Auto

As Pearl, with off/constant/auto control

Horstmann 424 Sapphire

For diagram see Figure 2.14, page 58

Electromechanical 24 hour priority programmer

On/off × 2
H130 × W87 × D57
Switch rating 6A (3A)

Horstmann 424 Topaz

For diagram see Figure 2.4

Electromechanical 24 hour time switch

Specifically designed for use with night setback thermostat

On/off × 2
H130 × W87 × D57
Switch rating 6A (3A)

Horstmann 425 Coronet

E	N	L	1	2	3	4	5	6
○	○	○	○	○	○	○	○	○
E	N	L		SPARE		ON	COM	OFF
	MAINS							

Voltage free switching unless L–5 linked

Electromechanical 24 hour time switch

On/off × 2
H107 × W152 × D39
Switch rating 16A (6A)

Horstmann 425 Diadem

E	N	L	1	2	3	4	5	6
○	○	○	○	○	○	○	○	○
E	N	L	HW	COM	HW	CH	COM	CH
	MAINS		ON		OFF	ON		OFF

Voltage free switching unless L–2–5 linked

Electromechanical 24 hour Basic/Full programmer

On/off × 2
H107 × W152 × D39
Switch rating 6A (2A)
See 425 TIARA

Horstmann 425 Tiara

FULLY PUMPED HOT WATER & CENTRAL HEATING

GRAVITY HOT WATER, PUMPED CENTRAL HEATING

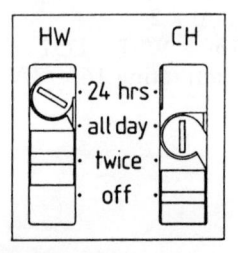

HW CH — 24 hrs / all day / twice / off

HW CH — 24 hrs / all day / twice / off

Figure 2.5

As 425 Diadem but without neon indicators

To set programmer for Basic/Full control, turn interlock screws as shown

Programmers and time switches

Horstmann 517

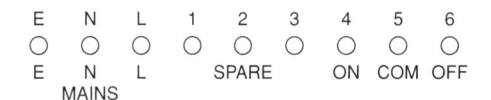

E	N	L	1	2	3	4	5	6
○	○	○	○	○	○	○	○	○
E	N	L		SPARE		ON	COM	OFF
MAINS								

Voltage free switching unless L–5 linked

Electronic 7 day time switch

On/off × 3
H101 × W175 × D45
Switch rating 3A (1A)

Horstmann 525

E	N	L	1	2	3	4	5	6
○	○	○	○	○	○	○	○	○
E	N	L	HW	COM	HW	CH	COM	CH
MAINS			ON		OFF	ON		OFF

Voltage free switching unless L–2–5 linked

Electronic 24 hour Basic/Full programmer

Facility for setting hot water and heating at different times to each other daily in Full mode

On/off × 2
H101 × W175 × D45
Switch rating 3A (1A)

For Full mode switch on power to the programmer, remove the switch over plate and move slide switch to extreme left. Move slide switch three positions to the right and re-fit switch cover plate. For Basic mode move the slide switch to the extreme right and fit gravity cover plate and switch on power.

Horstmann 525 7D

E	N	L	1	2	3	4	5	6
○	○	○	○	○	○	○	○	○
E	N	L	HW	COM	HW	CH	COM	CH
MAINS			ON		OFF	ON		OFF

Voltage free switching unless L–2–5 linked

Electronic 7 day Basic/Full programmer

Facility for setting hot water and heating at different times from each other every day in Full mode.

On/off × 2
H101 × W175 × D45
Switch rating 3A (1A)
Convert to Full/Basic mode as 525

Horstmann 525 Zone

As 527 7D but outputs labelled Zone 1 and Zone 2 instead of HW and CH.

Horstmann 581 Senior

Typical wiring diagram

Electronic 24 hour time switch

Designed for use by the elderly on systems utilizing combination boilers. It has a night set-back facility and is supplied with a suitable room thermostat (Eberle 3545) for which wiring instructions are given. The usual on/off times are featured as high/low.

High/low × 2
H101 × W175 × D45
Switch rating 3A (1A)

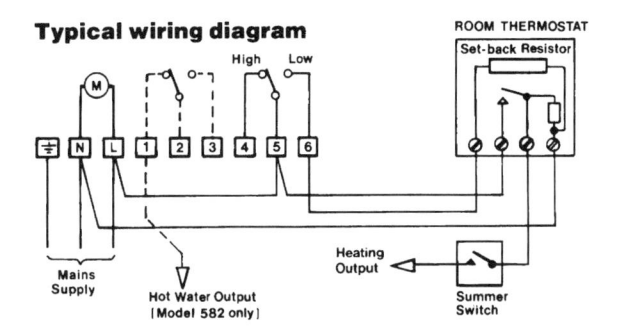

Figure 2.6

Horstmann 582 Senior

As 581 but with additional hot water switching
See Figure 2.6

Electronic 24 hour Basic/Full programmer

On/off × 2
H101 × W175 × D45
Switch rating 3A (1A)
Convert to Full/Basic as 525

Horstmann H 11

```
E   N   L   1     2    3   4    5    6
O   O   O   O     O    O   O    O    O
E   N   L       SPARE     ON  COM  OFF
  MAINS
```

Voltage free switching unless L–5 linked

Electronic 24 hour time switch

On/off × 3
H101 × W163 × D33
Switch rating 3A (1A)

Horstmann H 17

```
E   N   L   1     2    3   4    5    6
O   O   O   O     O    O   O    O    O
E   N   L       SPARE     ON  COM  OFF
  MAINS
```

Voltage free switching unless L–5 linked

Electronic 7 day time switch

On/off × 3
H101 × W163 × D33
Switch rating 3A (1A)

Horstmann H 21

```
E   N   L   1    2    3    4    5    6
O   O   O   O    O    O    O    O    O
E   N   L   HW  COM   HW   CH  COM   CH
  MAINS    ON        OFF   ON       OFF
```

Voltage free switching unless L–2–5 linked

Electronic 24 hour Basic/Full programmer

On/off × 3
H101 × W163 × D33
Switch rating 3A (1A)
To change from Basic to Full move slider at rear of programmer.

Horstmann H 27

```
E   N   L   1    2    3    4    5    6
O   O   O   O    O    O    O    O    O
E   N   L   HW  COM   HW   CH  COM   CH
  MAINS    ON        OFF   ON       OFF
```

Voltage free switching unless L–2–5 linked

Electronic 7 day Basic/Full programmer

Facility for setting hot water and heating at different times from each other every day in Full mode.

On/off × 3
H101 × W163 × D33
Switch rating 3A (1A)

Horstmann H 27 Z

As H27 but outputs labelled Zone 1 and Zone 2 instead of HW and CH

Programmers and time switches

Horstmann H 37

E	N	L						
○	○	○	○	○	○	○	○	○
MAINS 240V			ON ZONE 1	OFF	ON ZONE 2	OFF	ON ZONE 3	OFF

Although the programmer commissioning switch has a gravity position it should not be selected

Electronic 7 day Full programmer

With one hot water channel and two heating zone channels.

On/off × 3
H101 × W163 × D33
Switch rating 3A (1A)

Horstmann H 121

E	N	L	1	2	3	4	5	6
○	○	○	○	○	○	○	○	○
E	N	L	HW ON	COM	HW OFF	CH ON	COM	CH OFF
MAINS								

Voltage free switching unless L–2–5 linked

Electronic 24 hour Basic/Full programmer

On/off × 3
H101 × W163 × D33
Switch rating 3A (1A)

Horstmann SC1 Centaur

1	○	N
2	○	SPARE
3	○	ON
4	○	LIVE

The unit is battery powered and so no neutral is required

Electronic 24 hour time switch

On/off × 3
H71 × W142 × D30
Switch rating 5A (1A)

Horstmann SC 7 Centaur

5/2 day version of SC1 with same wiring and specification.

Horstmann TC 1 Centaur

1	○	N
2	○	CH ON
3	○	HW ON
4	○	LIVE IN

The unit is battery powered and so no neutral is required

Electronic 24 hour basic programmer

On/off × 3
H71 × W142 × D30
Switch rating 5A (1A)

Horstmann TC 7 Centaur

5/2 day version of TC 1 with same wiring and specification.

Horstmann KMK 2A

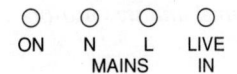

```
  O     O   O    O
  ON    N   L   LIVE
       MAINS    IN
```

Link L-Live In if required

Electromechanical 24 hour time switch

This is a plug-in unit, suitable for fitting into a terminal board or similar.

On/off × 2
H110 × W80 × D77
Switch rating 30A (20A)

Horstmann YMK 2

As KMK 2A with 12 hour spring reserve

Ideal STD. ISC-1

```
  N    L    1     2     3     4
  O    O    O     O     O     O
  N    L   SPARE  CH   SPARE  HW
  MAINS          ON          ON
```

Electromechanical 24 hour Basic/Full programmer

On/off × 2
H105 × W181 × D77
Switch rating 3A (1A)

International Janitor

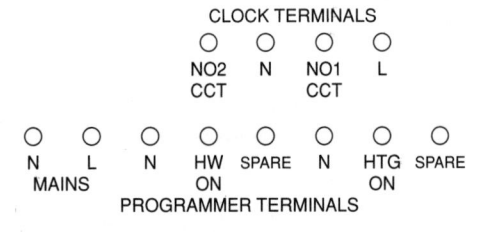

```
              CLOCK TERMINALS
               O    O    O    O
              NO2   N   NO1   L
              CCT        CCT

  O    O    O    O    O    O    O    O
  N    L    N    HW  SPARE  N  HTG  SPARE
 MAINS         ON              ON
           PROGRAMMER TERMINALS
```

Programmer and clock internally wired

24 hour basic programmer

On/off × 2
H238 × W114 × D95
Switch rating 4A

Landis & Gyr RWB 1

```
  N    L              3     4
  O    O              O     O
  N    L              HW    CH
 MAINS                ON    ON
```

Electromechanical 24 hour Basic/Full programmer

On/off × 2
H80 × W135 × D38
Switch rating 10A (2A)
For Full control turn screw at rear of programmer to horizontal

Landis & Gyr RWB 2

```
  N    L    1     2     3     4
  O    O    O     O     O     O
  N    L    HW    CH    HW    CH
 MAINS     OFF   OFF    ON    ON
```

Electromechanical 24 hour Basic/Full programmer

On/off × 2
H80 × W135 × D38
Switch rating 10A (2A)
For Full control turn screw at rear of programmer to horizontal

Landis & Gyr RWB 2 MK2

As RWB 2 with internal electronic operation.
Switch rating 5A (2A)

Landis & Gyr RWB 2.9

As RWB 2, but without neon indicators

Landis & Staefa RWB 7

N	L	1	2	3	4
N	L	SPARE	COM	OFF	ON
MAINS					

Voltage free switching unless L–2 linked

Electronic 24 hour, 5/2 day, 7 day time switch

On/off × 2
H85 × W140 × D35
Switch rating 6A (2A)

Landis & Staefa RWB 9

N	L	1	2	3	4
N	L	HW OFF	CH OFF	HW ON	CH ON
MAINS					

Electronic 24 hour, 5/2 day, 7 day Basic/Full programmer

On/off × 2
H85 × W140 × D35
Switch rating 6A (2A)
To change from Full to Basic move dip switch on rear of programmer to '10' position

Landis & Gyr RWB 20

N	L	1	2	3	4
	L	HW OFF	CH OFF	HW ON	CH ON
MAINS					

The unit is battery powered so no neutral is required. Facility for setting hot water and heating at different times daily in Full mode

Electronic 7 day Basic/Full programmer

On/off × 3
H87 × W135 × D40
Switch rating 6A (2A)
For Full control cut link at rear of programmer

Landis & Gyr RWB 30

N	L	1	2	3	4
N	L	SPARE	COM	OFF	ON
MAINS					

Voltage free switching unless L–2 linked

Electromechanical 24 hour time switch

On/off × 2
H80 × W135 × D38
Switch rating 6A (2A)

Landis & Gyr RWB 40

N	L	1	2	3	4
N	L	HW OFF	CH OFF	HW ON	CH ON
MAINS					

Electronic 24 hour Basic/Full programmer

On/off × 3
H90 × W115 × D44
Switch rating 6A (2A)
For Full control cut link at rear of programmer

Landis & Gyr RWB 50

```
N    L    1      2     3     4
O    O    O      O     O     O
N    L    SPARE  COM   OFF   ON
MAINS
```

Voltage free switching unless L–2 linked

Electronic 24 hour time switch

On/off × 3
H90 × W115 × D44
Switch rating 6A (2A)

Landis & Gyr RWB 100

```
N    L    1      2     3     4
O    O    O      O     O     O
N    L    SPARE  COM   OFF   ON
MAINS
```

Voltage free switching unless L–2 linked

Electronic 24 hour time switch

On/off × 2
H80 × W135 × D31
Switch rating 6A (2A)

Landis & Gyr RWB 102

```
N    L    1     2      3     4
O    O    O     O      O     O
N    L    HW    CH     HW    CH
MAINS     OFF   OFF    ON    ON
```

Note that heating only is not available and no connection need to be made to terminal 1 as, for example, for a mid-position valve

Electronic 24 hour Basic programmer

On/off × 2
H80 × W135 × D31
Switch rating 6A (2A)

Landis & Gyr RWB 152

```
N    L    1      2     3     4
O    O    O      O     O     O
N    L    SPARE  COM   OFF   ON
MAINS
```

Voltage free switching unless L–2 linked

Electronic 5/2 day time switch

On/off × 2
H80 × W135 × D31
Switch rating 6A (2A)

Landis & Gyr RWB 170

```
N    L    1      2     3     4
O    O    O      O     O     O
N    L    SPARE  COM   OFF   ON
MAINS
```

Voltage free switching unless L–2 linked

Electronic 7 day time switch

On/off × 2
H80 × W135 × D31
Switch rating 6A (2A)

Landis & Gyr RWB 200

```
N    L    1     2      3     4
O    O    O     O      O     O
N    L    HW    CH     HW    CH
MAINS     OFF   OFF    ON    ON
```

Electronic 24 hour Basic/Full programmer

On/off × 2
H80 × W135 × D31
Switch rating 6A (2A)

Landis & Gyr RWB 252

N	L	1	2	3	4
○	○	○	○	○	○
N	L	HW	CH	HW	CH
MAINS		OFF	OFF	ON	ON

Landis & Gyr RWB 270

N	L	1	2	3	4
○	○	○	○	○	○
N	L	HW	CH	HW	CH
MAINS		OFF	OFF	ON	ON

Myson Microtimer 1

N	○	N MAINS		
L	○	L MAINS		
7	○	HW OFF		
6	○	HW ON	○	9
4	○	CH OFF	○	10
3	○	CH ON	○	11
				SPARE

A connection block is provided for neutrals

Myson Microtimer 7

N	○	N MAINS		
L	○	L MAINS		
7	○	HW OFF		
6	○	HW ON	○	9
4	○	CH OFF	○	10
3	○	CH ON	○	11
				SPARE

A connection block is provided for neutrals

Potterton 411

E	1	2	3	4	5	6	7
○	○	○	○	○	○	○	○
E	N	L	HW	CH			SPARE
			ON	ON			
E	N	L	8	9	10	11	12	13
○	○	○	○	○	○	○	○	○
MAINS			N	L				

If fitting room stat connect across 6–11 and remove link. Internal links N–1–8, L–2–9, 3–10, 4–11, 6–13 are fitted

Electronic 5/2 day Basic/Full programmer

On/off × 2
H80 × W135 × D31
Switch rating 6A (2A)

Electronic 7 day programmer

On/off × 2
H80 × W135 × D31
Switch rating 6A (2A)

Electronic 24 hour Basic/Full programmer

On/off × 2
H100 × W165 × D50
Switch rating 2A (2A)
To adjust from Full to Basic system, remove programmer from backplate and move system select switch to required position.

Electronic 7 day Basic/Full programmer

On/off × 2
H100 × W165 × D50
Switch rating 2A (2A)
To adjust from Full to Basic system, remove programmer from backplate and move system select switch to required position.

Electromechanical 24 hour Basic programmer

On/off × 2
H193 × W103 × D82
Switch rating 5A

Potterton 420 Prefect

Programmer incorporated relay and was used in conjunction with a cylinder thermostat and hot water motor open/close motorized valve on a gravity hot water pumped central heating system.

Electromechanical 24 hour Full programmer

On/off × 2
H117 × W254 × D90
Switch rating 2A

Potterton 423

As Horstmann 423 Diamond

Potterton 424

As Horstmann 424 Diamond

Potterton EP 2000

```
A    B    C    D    N    L    1    2    3    4    5
O    O    O    O    O    O    O    O    O    O    O
          SPARE          MAINS     HW   CH   HW   CH   L
                                   OFF  OFF  ON   ON
```

Link L–5. A connection block is provided for neutrals and earths

Electronic 24 hour Basic/Full programmer

On/off × 2
H100 × W157 × D46
Switch rating 6A (2A)
To adjust from Basic to Full move slider from 10 to 16 position and turn screw to vertical on rear of programmer.

Potterton EP 2000 MK 2

```
A    B    C    D    N    L    1    2    3    4    5
O    O    O    O    O    O    O    O    O    O    O
          SPARE          MAINS     HW   CH   HW   CH   L
                                   OFF  OFF  ON   ON
```

Link L–5. A connection block is provided for neutrals and earths

Electronic 24 hour Basic/Full programmer

On/off × 2
H104 × W161 × D49
Switch rating 2 (1A)
To adjust from Basic to Full move slider on rear of programmer from 10 to 16 position.

Potterton EP 2001

```
A    B    C    D    N    L    1    2    3    4    5
O    O    O    O    O    O    O    O    O    O    O
          SPARE          MAINS     HW   CH   HW   CH   L
                                   OFF  OFF  ON   ON
```

Link L–5. A connection block is provided for neutrals and earths

Electronic 24 hour Basic/Full programmer

Facility for 5/2 day setting

On/off × 2
H104 × W160 × D41
Switch rating 6A (2A)
To adjust from Basic to Full move slider from 10 to 16 position on rear of programmer with battery removed.

Potterton EP 2002

Electronic 5/2 day Basic/Full programmer

5/2 day version of EP 2000 MK2

Potterton EP 3000

As EP 2000 specification but with 7 day programming facility

Programmers and time switches

Potterton EP 3001

As EP 2001 specification but with 7 day programming facility and allows for hot water and central heating to be set for different times daily

Potterton EP 3002

Electronic 7 day Basic/Full programmer

Facility for setting hot water and heating for different times daily.

7 day version of EP 2000 MK2

Potterton EP 4000

A	B	C	D	N	L	1	2	3	4	5
○	○	○	○	○	○	○	○	○	○	○
SPARE				MAINS		SP	OFF	SP	ON	COM

Voltage free switching unless L–5 linked. A connection block is provided for neutrals and earths.

Electronic 7 day time switch

On/off × 2
H100 × W157 × D46
Switch rating 6A (2A)

Potterton EP 4000 MK2

A	B	C	D	N	L	1	2	3	4	5
○	○	○	○	○	○	○	○	○	○	○
SPARE				MAINS		SP	OFF	SP	ON	COM

Voltage free switching unless L–5 linked. A connection block is provided for neutrals and earths.

Electronic 24 hour time switch

On/off × 2
H104 × W161 × D49
Switch rating 2 (1A)

Potterton EP 4001

A	B	C	D	N	L	1	2	3	4	5
○	○	○	○	○	○	○	○	○	○	○
SPARE				MAINS		SP	OFF	SP	ON	COM

Voltage free switching unless L–5 linked. A connection block is provided for neutrals and earths.

Electronic 5/2 day time switch

On/off × 3
H104 × W160 × D41
Switch rating 6A (2A)

Potterton EP 4002

Electronic 5/2 day time switch

5/2 day version of EP 4000 MK2 with 3 on/offs per day.

Installing and Servicing Domestic Central Heating Wiring Systems and Controls

Potterton EP 5001

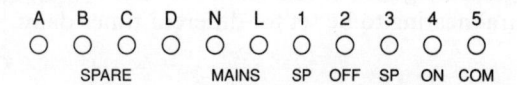

A	B	C	D	N	L	1	2	3	4	5
○	○	○	○	○	○	○	○	○	○	○
	SPARE			MAINS		SP OFF	SP ON	COM		

Voltage free switching unless L–5 linked. A connection block is provided for neutrals and earths.

Electronic 7 day time switch

On/off × 3
H104 × W160 × D41
Switch rating 6A (2A)

Potterton EP 5002

Electronic 7 day time switch

7 day version of EP 4000 MK2 with 3 on/offs per day.

Potterton EP 6000

A	B	C	D	N	L	1	2	3	4	5
○	○	○	○	○	○	○	○	○	○	○
	SPARE			MAINS		HW OFF	CH OFF	HW ON	CH ON	L

Link L–5. A connection block is provided for neutrals and earths.

Electronic 7 day Basic/Full programmer

Facility for setting either hot water or heating at different times daily and the other channel 5/2 day in Full mode.

On/off × 3
H104 × W164 × D51
Switch rating 6A (2A)
To adjust from Basic to Full move slider from 10 to 16 position on rear of programmer with battery removed.

Potterton EP 6002

A	B	C	D	N	L	1	2	3	4	5
○	○	○	○	○	○	○	○	○	○	○
	SPARE			MAINS		HW OFF	CH OFF	HW ON	CH ON	L

Link L–5. A connection block is provided for neutrals and earths.

Electronic 7 day Full programmer

Facility for setting hot water and heating for different times to each other daily

On/off × 3
H104 × W161 × D49
Switch rating 2 (1A)

Potterton Mini-Minder

As Landis & Gyr RWB 2

Potterton Mini-Minder E

N	L	1	2	3	4
○	○	○	○	○	○
MAINS 240V		HW OFF	CH OFF	HW ON	CH ON

Link L–5. A connection block is provided for neutrals and earths.

Electronic 24 hour Basic/Full programmer

On/off × 2
H105 × W164 × D51
Switch rating 2A (1A)
To change Basic/Full set both sliders to OFF and turn selector at rear to required position.

Programmers and time switches

Potterton Mini-Minder Es

N	L	1	2	3	4
○	○	○	○	○	○
MAINS 240V		SPARE	COM	OFF	ON

Voltage free switching unless L–2 linked

Electronic 24 hour time switch

On/off × 2
H105 × W164 × D51
Switch rating 2A (1A)

Proheat FP 1

As Flash 31031

Proheat FP 2

As Flash 31032

Proheat FP 3

As Flash 31033

Randall Mk. 1

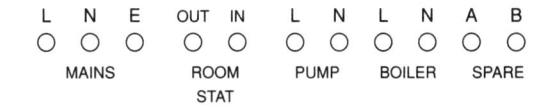

L	N	E	OUT	IN	L	N	L	N	A	B
○	○	○	○	○	○	○	○	○	○	○
MAINS			ROOM STAT		PUMP		BOILER		SPARE	

Electromechanical 24 hour Basic programmer

On/off × 2
H100 × W200 × D68

Randall Mk. 2 R6

1	2	3	4	5	6	7
○	○	○	○	○	○	○
L	L	N	HW ON	CH ON	SPARE	DO NOT USE

Link 1–2

Electromechanical 24 hour Basic programmer

On/off × 2
H216 × W102 × D57
Switch rating 5A

Randall 102

1	2	3	E	5	6
○	○	○	○	○	○
HW ON	CH ON	COM	E	N MAINS	L

Voltage free switching unless 3–6 linked

Electromechanical 24 hour Basic programmer

On/off × 2
H135 × W112 × D69
Switch Rating 6A

Randall 102 E

1	2	3	E	5	6
○	○	○	○	○	○
HW ON	CH ON	COM	E	N MAINS	L

Voltage free switching unless 3–6 linked

Electronic 24 hour Basic programmer

On/off × 6
H136 × W102 × D47
Switch rating 3A

Randall 102 E5

```
1     2     3     E     5     6
O     O     O     O     O     O
HW    CH    COM   E     N     L
ON    ON                MAINS
```

Voltage free switching unless 3–6 linked

Electronic 5/2 day Basic programmer

On/off × 3
H136 × W102 × D47
Switch rating 3A

Randall 102 E7

```
1     2     3     E     5     6
O     O     O     O     O     O
HW    CH    COM   E     N     L
ON    ON                MAINS
```

Voltage free switching unless 3–6 linked

Electronic 7 day Basic programmer

On/off × 3
H136 × W102 × D47
Switch rating 3A

Randall 103

```
1     2     3     E     5     6
O     O     O     O     O     O
ON    SPARE COM   E     N     L
                        MAINS
```

Voltage free switching unless 3–6 linked

Electromechanical 24 hour time switch

On/off × 2
H135 × W112 × D69
Switch Rating 6A

Randall 103 E

```
1     2     3     E     5     6
O     O     O     O     O     O
ON    SPARE COM   E     N     L
                        MAINS
```

Voltage free switching unless 3–6 linked

Electronic 24 hour time switch

On/off × 6
H136 × W102 × D47
Switch rating 3A

Randall 103 E5

```
1     2     3     E     5     6
O     O     O     O     O     O
ON    SPARE COM   E     N     L
                        MAINS
```

Voltage free switching unless 3–6 linked

Electronic 5/2 day time switch

On/off × 3
H136 × W102 × D47
Switch rating 3A

Randall 103 E7

```
1     2     3     E     5     6
O     O     O     O     O     O
ON    SPARE COM   E     N     L
                        MAINS
```

Voltage free switching unless 3–6 linked

Electronic 7 day time switch

On/off × 3
H136 × W102 × D47
Switch rating 3A

Programmers and time switches

Randall 105

For use with the ACL/Tower Biflo mid-position valve.

For diagram see Figure 2.24, page 61

Randall 106

For diagram see Figure 2.15, page 58

Randall 151

```
 1     2     3    E    5     6
 O     O     O    O    O     O
ON    OFF   COM   E    N     L
                     MAINS
```

Voltage free switching unless 3–6 linked

Randall 152 E

```
 1     2     3    E    5     6
 O     O     O    O    O     O
ON    OFF   COM   E    N     L
                     MAINS
```

Voltage free switching unless 3–6 linked

Randall 152 E7

```
 1     2     3    E    5     6
 O     O     O    O    O     O
ON    OFF   COM   E    N     L
                     MAINS
```

Voltage free switching unless 3–6 linked

Randall 153 E

```
 1     2     3    E    5     6
 O     O     O    O    O     O
ON   SPARE  COM   E    N     L
                     MAINS
```

Voltage free switching unless 3–6 linked

Electromechanical 24 hour Basic programmer

On/off × 2
H135 × W112 × D69
Switch rating 10A

Electromechanical 24 hour priority programmer

On/off × 2
H135 × W112 × D69
Switch rating 10A

Electromechanical 24 hour time switch

On/off × 2
H135 × W112 × D69
Switch rating 15A

Electronic 24 hour time switch

On/off × 6
H136 × W102 × D47
Switch rating 8A

Electronic 7 day time switch

On/off × 3
H136 × W102 × D47
Switch rating 8A

Electronic 24 hour time switch

On/off × 6
H136 × W102 × D47
Switch rating 15A (4A)

Randall 153 E7

1	2	3	E	5	6
○	○	○	○	○	○
ON	SPARE	COM	E	N	L
				MAINS	

Voltage free switching unless 3–6 linked

Electronic 7 day time switch

On/off × 3
H136 × W102 × D47
Switch rating 15A (4A)

Randall 701

1	2	3	4	5	6	L	N	E
○	○	○	○	○	○	○	○	○
CH ON	CH OFF	HW ON	HW OFF	L CH	L HW	MAINS		

Voltage free switching unless 5–6–L linked

Electronic 24 hour Basic programmer

On/off × 3
H108 × W221 × D51
Switch rating 3A

Randall 702

1	2	3	4	5	6	L	N	E
○	○	○	○	○	○	○	○	○
CH ON	CH OFF	HW ON	HW OFF	L CH	L HW	MAINS		

Voltage free switching unless 5–6–L linked

Electronic 24 hour Full programmer

On/off × 3
H108 × W221 × D51
Switch rating 3A

Randall 811, 841, 842, 851, 852

A range of electronic time switches and programmers intended for commercial use. However, as it is possible to use these in a domestic situation, basic wiring and description are given here.

All models:
On/off × up to 200 on/off operations in any 7 day period.
H112 × W226 × D55

Randall 811

	N	L	E	1	2
○	○	○	○	○	○
SPARE	MAINS			COM	ON

Voltage free switching unless L–1 linked

7 day time switch

With 30A SPST switch

Randall 851

E	N	L	6	5	4	3	2	1
○	○	○	○	○	○	○	○	○
SPARE	MAINS			SPARE		OFF	ON	COM

Voltage free switching unless L–1 linked

7 day time switch

With 10A SPDT switch

Programmers and time switches

Randall 852

E	N	L	6	5	4	3	2	1
○	○	○	○	○	○	○	○	○
SPARE	MAINS		OFF	ON	COM	OFF	ON	COM
				CHANNEL 2			CHANNEL 1	

Voltage free switching unless L–4–1 linked

2 channel 7 day time switch

With independent SPDT switching to each channel.

Randall 841

N	L	E	1	2
○	○	○	○	○
SPARE	MAINS		COM	ON

Voltage free switching unless L–1 linked

Single channel pulsed output time switch

For applications where a time pulsed signal is required to activate equipment. For example, for timed bell ringing to school class changes, or shift changes and breaks in factories.

Randall 842

E	N	L	6	5	4	3	2	1
○	○	○	○	○	○	○	○	○
SPARE	MAINS		OFF	ON	COM	OFF	ON	COM
				CHANNEL 2			CHANNEL 1	

Voltage free switching unless L–4–1 linked

2-channel pulsed output time switch

Each output channel is totally independent of the other and can be used where timed pulsed signals are required to activate equipment in two separate areas, for example, in locations as described above.

Randall 911

E	N	L	1	2	3	4	5	6
○	○	○	○	○	○	○	○	○
	MAINS			SPARE		OFF	COM	ON

Voltage free switching unless L–5 linked

Electronic 24 hour time switch

On/off × 6
H85 × W160 × D38
Switch rating 3A

Randall 922

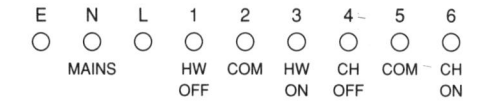

E	N	L	1	2	3	4	5	6
○	○	○	○	○	○	○	○	○
	MAINS		HW OFF	COM	HW ON	CH OFF	COM	CH ON

Voltage free switching unless L–2–5 linked

Facility for setting hot water and heating at different times from each other daily in Full mode.

Electronic 24 hour Basic/Full programmer

On/off × 6
H85 × W160 × D38
Switch rating 3A
For basic system the recessed switch on the rear of the programmer should be moved to the **up** position.

Randall 971

E	N	L	1	2	3	4	5	6
○	○	○	○	○	○	○	○	○
	MAINS			SPARE		OFF	COM	ON

Voltage free switching unless L–5 linked

Electronic 7 day time switch

On/off × 3
H85 × W160 × D38
Switch rating 3A

Randall 972

E	N	L	1	2	3	4	5	6
O	O	O	O	O	O	O	O	O
	MAINS		HW OFF	COM	HW ON	CH OFF	COM	CH ON

Voltage free switching unless L–2–5 linked

Facility for setting hot water and heating at different times from each other daily in Full mode.

Electronic 7 day Basic/Full programmer

On/off × 3
H85 × W160 × D38
Switch rating 3A

Randall 3020P

1	2	3	4	5	6	7	E
O	O	O	O	O	O	O	O
N	CH ON	SP	HW ON	SP	L	N MAINS	E

Electromechanical 24 hour Basic programmer

On/off × 2
H216 × W102 × D57
Switch rating 3A

Randall 3022

For diagram see Figure 2.17, page 59

Electromechanical 24 hour priority programmer

On/off × 2
H216 × W102 × D57
Switch rating 3A

Randall 3033

1	2	3	4	5	6	7	E
O	O	O	O	O	O	O	O
N	CH ON	CH OFF	HW ON	HW OFF	L	N MAINS	E

Electromechanical 24 hour Full programmer

On/off × 2
H216 × W102 × D57
Switch rating 3A

Randall 3060

1	2	3	4	5	6	7	E
O	O	O	O	O	O	O	O
N	CH ON	SP	HW ON	SP	L	N MAINS	E

Electromechanical 24 hour Basic programmer

On/off × 2
H216 × W102 × D57
Switch rating 3A

Randall 4033

1	2	3	4	5	6	7	E
O	O	O	O	O	O	O	O
L HW	CH ON	CH OFF	HW ON	HW OFF	L	N MAINS	E

Link 1–6

Electromechanical 24 hour Full programmer

On/off × 2
H216 × W102 × D57
Switch rating 3A

Programmers and time switches

Randall Set 1

E	N	L	1	2	3	4	5	6
○	○	○	○	○	○	○	○	○
MAINS				SPARE		ON	COM	OFF

Voltage free switching unless L–5 linked

Electronic 24 hour time switch

On/off × 2
H101 × W148 × D36
Switch rating 5A

Randall Set 2

E	N	L	1	2	3	4	5	6
○	○	○	○	○	○	○	○	○
MAINS			HW ON	COM	HW OFF	CH ON	COM	CH OFF

Voltage free switching unless L–2–5 linked

Electronic 24 hour Basic programmer

On/off × 2
H101 × W148 × D36
Switch rating 3A

Randall Set 3

E	N	L	1	2	3	4	5	6
○	○	○	○	○	○	○	○	○
MAINS			HW ON	COM	HW OFF	CH ON	COM	CH OFF

Voltage free switching unless L–2–5 linked

Electronic 24 hour Full programmer

On/off × 2
H101 × W148 × D36
Switch rating 3A

Randal Set 1E, 2E, 3E, 3M

See Danfoss Randall 1E, 2E, 3E, 3M

Randall Set 4

E	N	L	1	2	3	4	5	6
○	○	○	○	○	○	○	○	○
MAINS				SPARE		ON	COM	OFF

Voltage free switching unless L–5 linked

Electronic 7 day time switch

On/off × 2
H101 × W148 × D36
Switch rating 5A

Randall Set 5

E	N	L	1	2	3	4	5	6
○	○	○	○	○	○	○	○	○
MAINS			HW ON	COM	HW OFF	CH ON	COM	CH OFF

Voltage free switching unless L–2–5 linked

Electronic 5/2 day Full programmer

On/off × 2
H101 × W148 × D36
Switch rating 3A

Randall TSR/2

1	2	3	4	5	6	7
○	○	○	○	○	○	○
L	L	N	SPARE	ON		SPARE

Link 1–2

Electromechanical 24 hour time switch

On/off × 2
H216 × W102 × D57
Switch rating 3A

Randall TSR 2+2

For diagram see Figure 2.16, page 59

Electromechanical 24 hour priority programmer

On/off × 2
H216 × W102 × D57
Switch rating 3A

Randall TSR 2P

1	2	3	4	5	6	7
O	O	O	O	O	O	O
L	L	N	SPARE	HW ON		CH ON

Link 1–2 and 5–6

Electromechanical 24 hour Basic programmer

On/off × 2
H216 × W102 × D57
Switch rating 3A

Randall TSR 3+3

1	2	3	4	5	6	7
O	O	O	O	O	O	O
HW ON	HW OFF	N	CH ON	CH OFF	N	L MAINS

Electromechanical 24 hour Full programmer

On/off × 2
H216 × W102 × D57
Switch rating 3A

Ravenheat

As Switchmaster 950

Sangamo M5

8	O	CH ON
7	O	
6	O	CH L
5	O	N
4	O	N MAINS
3	O	L MAINS
2	O	HW OFF
1	O	HW ON

Link 1–6

Electromechanical 24 hour Basic programmer

On/off × 2
H89 × W141 × D38
Switch rating 10A (2A)

Sangamo M6

8	O	SPARE
7	O	SPARE
6	O	L MAINS
5	O	N
4	O	N MAINS
3	O	COM
2	O	OFF
1	O	ON

Voltage free switching unless 3–6 linked

Electromechanical 24 hour time switch

On/off × 2
H89 × W141 × D38
Switch rating 10A (2A)

Programmers and time switches

Sangamo S250 series and S350 series

```
3 TERMINAL           4 TERMINAL
O   O   O        O    O    O   O
L   N   ON       COM  ON   L   N
                                MAINS
```

Link L–COM if required

Electromechanical 24 hour time switch

On/off × 4 max.
H140 × W97 × D102
Switch rating 20A
Conduit box available, FD 930

Sangamo S408 Form 2

```
            O    ON
            O    COM
    N   O
    N   O        MAINS
    L   O        MAINS
```

Voltage free switching unless L–COM linked

Electromechanical 24 hour time switch

On/off × pegs
H105 × W105 × D60
Switch rating 15A

Sangamo S408 Form 5

```
    O    ON
    O    N
    O    L
```

Electromechanical 24 hour time switch

On/off × pegs
H105 × W105 × D60
Switch rating 15A

Sangamo S408 Form 6

```
    O    ON
    O    COM
    O    N MAINS
    O    L MAINS
```

Voltage free switching unless L–COM linked

Electromechanical 24 hour time switch

On/off × pegs
H105 × W105 × D60
Switch rating 15A

Sangamo 409 F1

```
1  O  N          N  O  MAINS
2  O  HW ON      L  O  MAINS
3  O  ON         5  O  ROOM STAT OUT
4  O  CH ON      6  O  ROOM STAT IN
```

If no room stat link 5–6

Electromechanical 24 hour Basic programmer

On/off × 2
H106 × W160 × D67
Switch rating 10A (3A)

Sangamo 409 F3

```
                 1  O  CH ON
3  O  N          2  O  CH OFF
6  O  N MAINS    4  O  HW ON
7  O  L MAINS    5  O  HW OFF
```

Electromechanical 24 hour Full programmer

On/off × 2
H106 × W160 × D67
Switch rating 10A (3A)

39

Sangamo 409 F4

1	○ N	N	○	MAINS
2	○ HW ON	L	○	MAINS
3	○ ON	5	○	ROOM STAT OUT
4	○ CH ON	6	○	ROOM STAT IN

If no room stat link 5–6

Electromechanical 24 hour Basic programmer

On/off × 2
H106 × W160 × D67
Switch rating 10A (3A)

Sangamo 409 F5

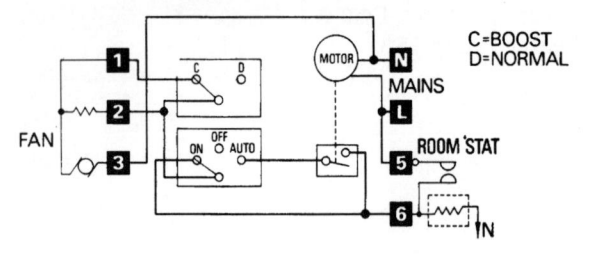

Figure 2.7

Electromechanical 24 hour Electricaire control

On/off × 2
H106 × W160 × D67
Switch rating 10A (3A)

Sangamo 409 F6

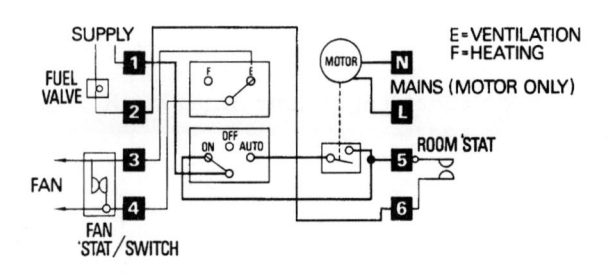

Figure 2.8

Electromechanical 24 hour warm air control

with provision for ventilation.

On/off × 2
H106 × W160 × D67
Switch rating 10A (3A)

Sangamo 409 F7

For diagram see Figure 2.18, page 59

Electromechanical 24 hour priority programmer

On/off × 2
H106 × W160 × D67
Switch rating 10A (3A)

Sangamo 409 F8

1	○ SPARE	N	○	N MAINS
2	○ ON	L	○	L MAINS
3	○ N	5	○	ROOM STAT OUT
		6	○	ROOM STAT IN

If no room stat link 5–6

Electromechanical 24 hour time switch

On/off × 2
H106 × W160 × D67
Switch rating 10A (3A)

Sangamo 410 F1

8	○	CH ON
7	○	CH OFF
6	○	CH COM
5	○	N
4	○	N MAINS
3	○	L MAINS
2	○	HW OFF
1	○	HW ON

Link 3–6

Electromechanical 24 hour Full programmer

On/off × 2
H85 × W138 × D46
Switch rating 10A (2A)

Sangamo 410 F2

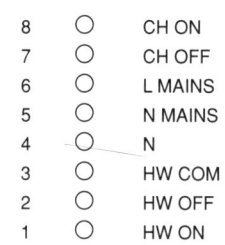

8	○	CH ON
7	○	CH OFF
6	○	L MAINS
5	○	N MAINS
4	○	N
3	○	HW COM
2	○	HW OFF
1	○	HW ON

Link 3–6

Electromechanical 24 hour Basic programmer

On/off × 2
H85 × W138 × D46
Switch rating 10A (2A)

Sangamo 410 F3

Electromechanical 24 hour programmer

Specification as 410 F1 but labelled Zone 1 and Zone 2 instead of hot water and central heating.

Sangamo 410 F4 (early model)

8	○	CH ON
7	○	
6	○	
5	○	N
4	○	N MAINS
3	○	L MAINS
2	○	
1	○	HW ON

Link 1–6

Electromechanical 24 hour Basic programmer

On/off × 2
H85 × W138 × D46
Switch rating 10A (2A)

Sangamo 410 F4

8	O	CH ON
7	O	CH OFF
6	O	CH COM
5	O	N
4	O	N MAINS
3	O	L MAINS
2	O	HW OFF
1	O	HW ON

Link 3–6

Electromechanical 24 hour Basic programmer

On/off × 2
H85 × W138 × D46
Switch rating 10A (2A)

Sangamo 410 F5

Electromechanical 24 hour Electricaire control

A single circuit programmer for controlling a two-speed fan on 'Electricaire' systems. It is fitted with an advance knob. The service knob allows a choice between 'Normal' or 'Boost' conditions.

On/off × 2
H85 × W138 × D46
Switch rating 10A (2A)

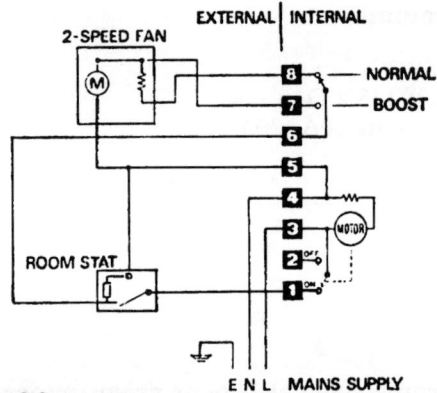

Figure 2.9

Sangamo 410 F6

Electromechanical 24 hour warm air control

A single circuit programmer for controlling warm air systems with a provision for VENT (ventilation). It is fitted with an advance knob, programme selector knob and a manual selector knob to provide VENT 'ON' or 'OFF'.

On/off × 2
H85 × W138 × D46
Switch rating 10A (2A)

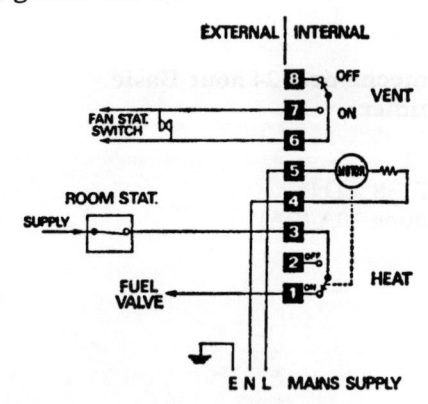

Figure 2.10

Sangamo 410 F7

For diagram see Figure 2.19, page 59

Electromechanical 24 hour priority programmer

On/off × 2
H85 × W138 × D46
Switch rating 10A (2A)

Programmers and time switches

Sangamo 410 F8

8	○	SPARE
7	○	SPARE
6	○	SPARE
5	○	N
4	○	N MAINS
3	○	L MAINS
2	○	OFF
1	○	ON

Electromechanical 24 hour time switch

On/off × 2
H85 × W138 × D46
Switch rating 10A (2A)

Sangamo 410 F9

For diagram see Figure 2.25, page 61

Electromechanical 24 hour Full programmer

For fully pumped systems utilizing ACL/Tower Biflo mid-position valve

On/off × 2
H85 × W138 × D46
Switch rating 10A (2A)

Sangamo 414 twin set

8	○	ZONE 1 ON
7	○	ZONE 1 OFF
6	○	L MAINS
5	○	N MAINS
4	○	N
3	○	ZONE 2 COM
2	○	ZONE 2 OFF
1	○	ZONE 2 ON

Link 3–6

Electromechanical 24 hour two-zone two-clock programmer

The programmer is fitted with two time switch dials on the left-hand side, each having two 'on' and two 'off' levers. The right-hand side is divided into two, the upper portion carrying the knobs and indicator associated with the upper time switch dial (Zone 1) and the lower portion with the lower dial (Zone 2).

On/off × 2 ea.
H85 × W138 × D61
Switch rating 10A (2A)

Sangamo 440

1	2	3	4	5	6	7	8
○	○	○	○	○	○	○	○
ON	SP	COM	N MAINS	SP	L MAINS	SP	SP

Voltage free switching unless 3–6 linked

Electromechanical 24 hour time switch

On/off × 2
H85 × W138 × D46
Switch rating 10A (2A)

Sangamo 931091

N	N	L	1	2
○—	—○	○	○	○
	MAINS		ON	COM

Voltage free switching unless L–2 linked

Electromechanical 24 hour time switch

On/off × multiple
H84 × W167 × D44
Switch rating 6A

Sangamo 931092

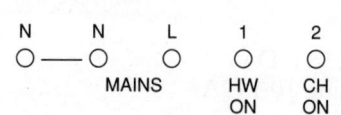

```
N    N    L    1    2
O────O    O    O    O
     MAINS  HW   CH
            ON   ON
```

Electromechanical 24 hour Basic programmer

On/off × multiple
H84 × W167 × D44
Switch rating 6A

Sangamo 931093

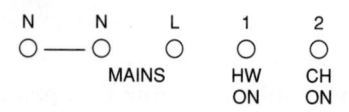

```
N    N    L    1    2
O────O    O    O    O
     MAINS  HW   CH
            ON   ON
```

Electromechanical 24 hour Full programmer

On/off × multiple
H84 × W167 × D44
Switch rating 6A

Sangamo S610

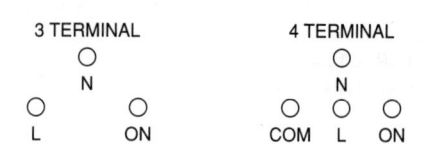

```
3 TERMINAL           4 TERMINAL
     O                    O
     N                    N
O         O         O    O    O
L         ON        COM  L    ON
```

Link L–COM if required

Electromechanical 24 hour time switch

On/off × 4
H139 × W82 × D70
Switch rating 30A
Conduit box available – FD 1510

Sangamo S611

As S610 with day omittance device

Sangamo Set 1

As Randall Set 1

Sangamo Set 2

As Randall Set 2

SMC   For diagrams see Figures 2.16 and 2.17, page 61; also Figures 11.48 and 11.49, page 208

The SMC programmers used in the SMC control pack (one boiler and two pumps, room and cylinder thermostats but no motorized valves) were manufactured by Horstmann. The first one was based on the 423 Pearl Auto and the next one was based on the 424 Emerald. They both contained a boiler switching relay. To replace, it is necessary to use either the new SMC wiring centre, incorporating a relay, or use a Basic or Full facility programmer and an external relay.

Smiths MK1 Centroller

```
P1   O         SPARE
P2   O         ON
P3   O         COM
     O    L
     O    N    MAINS 240V
     O    E
```

Electromechanical 24 hour time switch

On/off × 2
H140 × W99 × D57

Voltage free switching unless L–P3 linked

Programmers and time switches

Smiths MK2 Centroller

Electromechanical 24 hour time switch

As MK1 with on/off/auto control.

Smiths MK3 Centroller

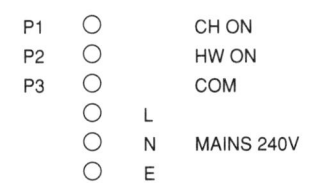

```
P1  O      CH ON
P2  O      HW ON
P3  O      COM
    O   L
    O   N  MAINS 240V
    O   E
```

Link L–P3

Electromechanical 24 hour basic programmer

On/off × 2
H140 × W99 × D57

Smiths Centroller 10

```
N   N   L   1    2    3    4
O   O   O   O    O    O    O
   MAINS       CH   HW
               ON   ON
```

Terminals 1 and 4 are internally linked and have no other connection

Electromechanical 24 hour Basic programmer

On/off × 2
H130 × W186 × D76
Switch rating 6A

Smiths Centroller 30

```
1    2    3      4    5    6
O    O    O      O    O    O
N    L   SPARE  ON   ON  SPARE
MAINS
```

Terminals 4 and 5 are internally linked

Electromechanical 24 hour time switch

On/off × 2
H98 × W156 × D72
Switch rating 6A

Smiths Centroller 30+

For diagram see Figure 2.20, page 60

Electromechanical 24 hour priority programmer

On/off × 2
H98 × W156 × D72
Switch rating 6A

Smiths Centroller 40

```
1    2    3      4    5
O    O    O      O    O
N    L   SPARE  HW   CH
MAINS            ON   ON
```

Electromechanical 24 hour Basic programmer

On/off × 2
H105 × W152 × D80
Switch rating 6A

Smiths Controller 40+

For diagram see Figure 2.21, page 60

Electromechanical 24 hour priority programmer

On/off × 2
H98 × W156 × D72
Switch rating 6A

Smiths Controller 50

```
  1   2   3   4   5
  O   O   O   O   O
  N   L  COM  ON  N
  MAINS
```

Voltage free switching unless 2–3 linked

Electromechanical 24 hour time switch

On/off × 2
H105 × W152 × D72
Switch rating 6A

Smiths Controller 60

```
  1   2    3     4    5
  O   O    O     O    O
  N   L  SPARE  HW   CH
  MAINS       ON   ON
```

Electromechanical 24 hour Basic programmer

On/off × 2
H105 × W152 × D80
Switch rating 6A

Smiths Controller 70

```
  1   2    3    4    5    6
  O   O    O    O    O    O
  N   L  SPARE CH   HW  SPARE
  MAINS        ON   ON
```

Electromechanical 24 hour Basic programmer

On/off × 2
H98 × W156 × D72
Switch Rating 6A

Smiths Controller 90

```
  1   2    3    4    5    6
  O   O    O    O    O    O
  N   L  SPARE CH   HW  SPARE
  MAINS        ON   ON
```

Electromechanical 24 hour Full programmer

On/off × 2
H98 × W156 × D72
Switch rating 6A

Smiths Controller 100

```
  N   N   L    1     2    3    4
  O   O   O    O     O    O    O
      N   L  SPARE  CH   HW  SPARE
      MAINS        ON   ON
```

Electromechanical 24 hour Basic programmer

On/off × 2
H130 × W186 × D76
Switch rating 6A

Smiths Controller 1000

```
  N   L    1     2     3    4
  O   O    O     O     O    O
  MAINS   HW    CH    HW   CH
          OFF   OFF   ON   ON
```

Electronic 24 hour Basic/Full programmer

On/off × 2
H84 × W140 × D40
Switch rating 2A

Smiths Centroller 2000
3000

See **Programmers and time switches with inbuilt or external sensors or thermostats** (Chapter 3).

Smiths Supply Master CHP11

Incorporated in a switched fused spur

6	5	4	3	2	1
O	O	O	O	O	O
L	N	HW	N	SPARE	CH
MAINS		ON			ON

Electronic 24 hour Basic programmer

On/off × 4
H85 × W85 × D37
Switch rating 3A

Smiths Supply Master CHP17

7 day version of CHP11

Smiths Supply Master FST11

Incorporated in a switched fused spur

6	5	4	3	2	1
O	O	O	O	O	O
L	N	N	L	COM	ON
MAINS					

Voltage free switching unless 3–2 linked

Electronic 24 hour time switch

On/off × 4
H85 × W85 × D37
Switch rating 13A

Smiths Supply Master FST17

7 day version of FST11

Sopac 24 Hour Fuelminder

1	O	CH ON
2	O	N MAINS
3	O	L MAINS
4	O	HW ON
5	O	DO NOT USE

Electromechanical 24 hour Basic programmer

On/off × 48
H159 × W83 × D42
Switch rating 15A (2A)

Sopac 7D Fuelminder

As 24 hour but with 7 day setting facility with on 6 on/offs per day.

Southern Digital

A range of time switches and controls not specifically designed for central heating but larger loads, e.g. immersion heaters.

Sugg Supaheat

1	2	3	4	5	6	7
O	O	O	O	O	O	O
CH	CH	N	HW	HW	N	L
ON	OFF		OFF	ON	MAINS	

Electromechanical 24 hour Full programmer

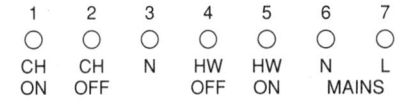

On/off × 2

47

Sunvic CB2201

Electromechanical 24 hour Full programmer

This unit is only available as a Clockbox 2 spare.

Sunvic DHP 1201 Libra

```
 1    2    3    4    5    6    7    8
 O    O    O    O    O    O    O    O
 N    L    CH   COM  CH   HW   COM  HW
 MAINS     ON        OFF  ON        OFF
```

Voltage free switching unless 2–4–7 linked

Electromechanical 24 hour Basic programmer

On/off × 6
H125 × W190 × D64
Switch rating 3A

Sunvic DHP 2201

```
 1    2    3    4    5    6    7    8
 O    O    O    O    O    O    O    O
 N    L    CH   COM  CH   HW   COM  HW
 MAINS     ON        OFF  ON        OFF
```

Voltage free switching unless 2–4–7 linked

Electronic 24 hour Full programmer

On/off × 6
H125 × W190 × D64
Switch rating 3A

Sunvic ET 1401/ET 1402

```
 1    2    3    4    5    6    7    8
 O    O    O    O    O    O    O    O
 N    L    COM  CH   CH   COM  HW   HW
 MAINS          ON   OFF       ON   OFF
```

Voltage free switching unless 2–3–6 linked

Electronic 24 hour Full programmer

On/off × 4
H110 × W180 × D65
Switch rating 5A (1A)

Sunvic ET 1451

As ET 1401 with battery reserve and advance features.

Sunvic Select 107

```
 N    L    1    2    3    4
 O    O    O    O    O    O
 MAINS     COM  OFF  ON   SPARE
```

Voltage free switching unless L–1 linked

Electronic 24 hour, 5/2 day, 7 day time switch

On/off × 2
H82 × W135 × D36
Switch rating 3A (1A)

Sunvic Select 207

```
 N    L    1    2    3    4
 O    O    O    O    O    O
 N    L    HW   CH   HW   CH
 MAINS     OFF  OFF  ON   ON
```

Electronic 24 hour, 5/2 day, 7 day Basic/Full programmer

On/off × 2
H82 × W135 × D36
Switch rating 3A (1A)

Programmers and time switches

Sunvic SP 20

```
     L   N       3    4     5
     O   O       O    O     O
     MAINS      COM SPARE  ON
```

Voltage free switching unless L–3 linked

Electronic 24 hour time switch

On/off × 2
H91 × W161 × D42
Switch rating 5A (1A)

Sunvic SP 25

```
 1   2   L   N   E   S   S   3   4   5
 O   O   O   O   O   O   O   O   O   O
 HW  HW     MAINS    SPARE    NOT USED  CH
 OFF ON                                 ON
```

Electronic 24 hour Basic programmer

On/off × 2
H91 × W161 × D42
Switch rating 5A (1A)

Sunvic SP 30

```
 1   2   L   N   E   S   S   3   4   5
 O   O   O   O   O   O   O   O   O   O
 HW  HW     MAINS    SPARE    NOT USED  CH
 OFF ON                                 ON
```

Electronic 24 hour Basic programmer

This unit is similar to the SP25 with extra features which may be useful for the elderly or visually handicapped, including setting tones.

On/off × 2
H91 × W161 × D42
Switch rating 5A (1A)

Sunvic SP 35

```
     L   N       3    4     5
     O   O       O    O     O
     MAINS      COM SPARE  ON
```

Voltage free switching unless L–3 linked

Electronic 24 hour time switch

On/off × 3
H91 × W161 × D42
Switch rating 6A (1A)

Sunvic SP 50

```
 1   2   L   N   E   S   S   3   4   5
 O   O   O   O   O   O   O   O   O   O
 HW  HW  L   N   E      SPARE   CH  CH  CH
 OFF ON     MAINS               COM OFF ON
```

Link L–3

Facility for setting hot water and heating at different times weekly in Full mode.

Electronic 24 hour Basic/Full programmer

On/off × 2
H91 × W161 × D42
Switch rating 6A (1A)
For Basic control cut link at rear of programmer.

Installing and Servicing Domestic Central Heating Wiring Systems and Controls

Sunvic SP 100

1	2	L	N	E	S	3	4	5
○	○	○	○	○	○	○	○	○
HW OFF	HW ON	L MAINS	N	E	SPARE	CH COM	CH OFF	CH ON

Link L–3

Facility for setting hot water and heating at different times to each other every day in Full mode.

Electronic 7 day Basic/Full programmer

On/off × 2
H91 × W161 × D42
Switch rating 6A (1A)
For Basic control cut link at rear of programmer.

Superswitch 1511

As Sopac Fuelminder 24H

Superswitch 1517

As Sopac Fuelminder 7D

Superswitch 1647

L	N	N	L1	L2
○	○	○	○	○
MAINS		LOAD	ON	SPARE

Electronic 24 hour time switch

On/off × 6
H90 × W162 × D75
Switch rating 15A (2)A

Superswitch 1657

L	N	N	L1	L2
○	○	○	○	○
MAINS		LOAD	HW ON	CH ON

Electronic 24 hour time switch

On/off × 6
H90 × W162 × D75
Switch rating 15A (2)A

Switchmaster 300

N	L	1	2	3	4
○	○	○	○	○	○
MAINS		ON	N	SPARE	COM

Links were fitted N–2 and L–4 but may have been removed depending on use. On the later ACL/Switchmaster version no links were fitted. Voltage free switch.

Electromechanical 24 hour time switch

On/off × 2
H100 × W168 × D48
Switch rating 5A (2A)

Switchmaster 320

N	L	1	2	3	4
○	○	○	○	○	○
MAINS		CH ON	N	HW ON	L

Links were fitted N–2 and L–4 on backplate

Electromechanical 24 hour Basic programmer

On/off × 2
H100 × W168 × D48
Switch rating 5A (2A)

Programmers and time switches

Switchmaster 350

N	L	1	2	3	4
○	○	○	○	○	○
MAINS		CH ON	N	HW ON	L

Links were fitted N–2 and L–4 on backplate

Electromechanical 24 hour Basic programmer

On/off × 2
H100 × W168 × D48
Switch rating 5A (2A)

Switchmaster 400

N	L	1	2	3	4
○	○	○	○	○	○
MAINS		CH ON	SPARE	HW ON	HW OFF

Electromechanical 24 hour Basic programmer

On/off × 2
H100 × W168 × D48
Switch rating 5A (2A)

Switchmaster 500

For diagram see Figure 2.22, page 60

Electromechanical 24 hour priority programmer

On/off × 2
H100 × W168 × D48
Switch rating 5A (2A)

Switchmaster 600

N	L	1	2	3	4
○	○	○	○	○	○
MAINS		CH ON	SPARE	HW ON	SPARE

Electromechanical 24 hour Basic programmer

On/off × 2
H100 × W168 × D48
Switch rating 5A (2A)

Switchmaster 800

As 805 with neon indicators.

Switchmaster 805

N	L	1	2	3	4
○	○	○	○	○	○
MAINS		CH ON	CH OFF	HW ON	HW OFF

Electromechanical 24 hour Full programmer

On/off × 2
H100 × W168 × D48
Switch rating 5A (2A)

Switchmaster 900

N	L	1	2	A	B	C	3	4
○	○	○	○	○	○	○	○	○
MAINS		CH ON	CH OFF		SPARE		HW ON	HW OFF

Electromechanical 24 hour Basic/Full programmer

On/off × 2
H138 × W83 × D53
Switch rating 5A (2A)
To adjust from Basic to Full mode, turn screw at rear of programmer from 10 to 16.

Switchmaster 905

As 900 with different styling.

51

Switchmaster 950

N	L	1	2	A	B	C
O	O	O	O	O	O	O
MAINS		ON	OFF		SPARE	

Electromechanical 24 hour time switch

On/off × 2
H138 × W83 × D53
Switch rating 5A (2A)

Switchmaster 980 Combi

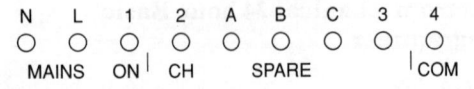

N	L	1	2	A	B	C	3	4
O	O	O	O	O	O	O	O	O
MAINS	ON	CH		SPARE				COM

Voltage free switching unless L–4 linked

Electromechanical 24 hour time switch

On/off × 2
H138 × W83 × D53
Switch rating 5A (2A)

Switchmaster 3000

As 300 with Homewarm motif.

Switchmaster 9000

N	L	1	2	A	B	C	3	4
O	O	O	O	O	O	O	O	O
MAINS		CH ON	CH OFF		SPARE		HW ON	HW OFF

Electronic 24 hour Basic/Full programmer

On/off × 2
H136 × W83 × D454
Switch rating 3A (2A)
To adjust from Basic to Full mode, move the slider at the rear of the programmer to the left.

Switchmaster 9001

As 9000 with different styling.

Switchmaster Sonata

L	N	1	2	4	6
O	O	O	O	O	O
MAINS		HW ON	HW OFF	CH ON	CH OFF

Electronic 7 day Full programmer

On/off × 3
H95 × W161 × D40
Switch rating 3A (2A)

Thorn

A version of the Randall 4033 was marketed with a Thorn cover and mounted horizontally.

Thorn Microtimer

As Honeywell ST 699 B.

Tower DP 72

N	L	1	2	3	4
O	O	O	O	O	O
MAINS		HW OFF	CH OFF	HW ON	CH ON

Facility for setting hot water and heating at different times for each other every day in Full mode.

Electronic 7 day Basic/Full programmer

On/off × 2
H79 × W163 × D50
Switch rating 5A (2A)
For Full mode remove tab switch at rear of programmer.

Programmers and time switches

Tower DT 71

N	L	1	2	3	4
○	○	○	○	○	○
MAINS		SPARE	OFF	COM	ON

Voltage free switching unless L–3 linked

Electronic 7 day time switch

On/off × 2
H79 × W163 × D50
Switch rating 5A (2A)

Tower FP

As ACL FP

Tower MP

As ACL MP

Tower QE1

N	L	1	2	3	4
○	○	○	○	○	○
MAINS		SPARE	OFF	COM	ON

Voltage free switching unless L–3 linked

Electronic 7 day time switch

On/off × 2
H85 × W155 × D45
Switch rating 5A (2A)

Tower QE2

N	L	1	2	3	4
○	○	○	○	○	○
MAINS		HW OFF	CH OFF	HW ON	CH ON

Electronic 7 day Basic/Full programmer

On/off × 2
H85 × W155 × D45
Switch rating 5A (2A)
To change from Basic to Full move slider at rear of programmer.

Tower TC

As ACL TC

Tower T2000

N	L		C	ON	OFF
○	○	○	○	○	○
MAINS			COM	CH ON	CH OFF

		N	C	ON	OFF
○	○	○	○	○	○
	SPARE	N	COM	HW ON	HW OFF

Voltage free switching unless L–C–C linked

Electronic 24 hour Full programmer

With facility for 5/2 day setting.

On/off × 2
H75 × W156 × D50
Switch rating 6A (2A)

53

Tower T2001

```
 N   L        7
 ○   ○   ○   ○   ○   ○
MAINS        ON      SPARE

 N   4
 ○   ○
 N  COM
```

Electromechanical 24 hour time switch

On/off × 4
H80 × W165 × D55
Switch rating 6A (2A)

Voltage free switching unless L–4 linked

Tower T2002

```
 N   L        N   N
 ○   ○   ○   ○   ○   ○
MAINS

 ○   ○   ○   ○   ○   ○
    SPARE    CH   HW
             ON   ON
```

Electromechanical 24 hour Basic programmer

On/off × 4
H80 × W165 × D55
Switch rating 6A (2A)

Tower T2003

```
 N   L        C   ON  OFF
 ○   ○   ○   ○   ○   ○
MAINS        COM  CH   CH
                  ON  OFF

              N   C   ON  OFF
 ○   ○   ○   ○   ○   ○
    SPARE    N  COM   HW   HW
                      ON  OFF
```

Electromechanical 24 hour Basic/Full programmer

On/off × 4
H80 × W165 × D55
Switch rating 6A (2A)
For Full mode remove plastic switch slide covers.

Voltage free switching unless L–C–C linked

Trac TP 10

As Flash 31031

Trac TP 20

As Flash 31032

Trac TP 20/7

As Flash 31732

Trac TP 30

As Flash 31033

Trac TP 30E

As Switchmaster 9001

Programmers and time switches

Venner CHC/2

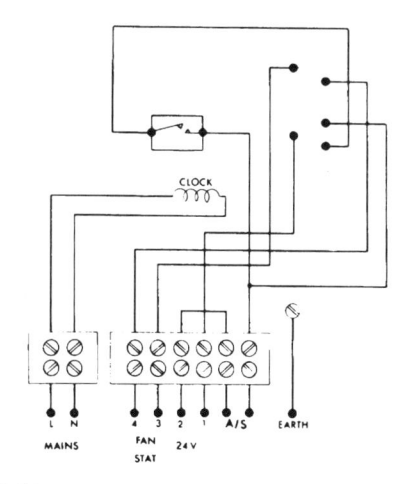

Figure 2.11

Electromechanical 24 hour warm air control

On/off × 2
H84 × W135 × D88
Switch rating 5A (2A)

Venner CHC/GW

For diagram see Figure 2.23, page 60

Electromechanical 24 hour priority programmer

On/off × 2
H84 × W135 × D88
Switch rating 5A (2A)

Venner CHC/W2

```
  AS        4    3    2    1    N    L
  O  O      O    O    O    O    O    O
  ROOM      N   CH    N   HW   MAINS
  STAT          ON        ON
```

Electromechanical 24 hour Basic programmer

On/off × 2
H84 × W135 × D88
Switch rating 5A (2A)

Venner T10

```
  3    2    1      N    L1   L    E
  O    O    O      O    O    O    O
  ON  COM  OFF    MAINS     MAINS
```

Voltage free switching unless L1–2 linked

Electromechanical 24 hour time switch

On/off × 2
H162 × W100 × D63
Switch rating 15A

Venner T20

```
  6    5    4    3    2    1    E    N    L
  O    O         O    O    O    O    O    O
  ROOM  CH    N    N   HW   E   MAINS
  STAT  ON             ON
```

Link 5–6 if no room stat

Electromechanical 24 hour Basic programmer

On/off × 2
H162 × W100 × D63
Switch rating 6A

Venner T30

6	5	4	3	2	1	E	N	L
○	○		○	○	○	○	○	○
ROOM STAT		CH ON	CH OFF	HW ON	HW OFF	E	MAINS	

Link 5–6 if no room stat

Electromechanical 24 hour Full programmer

On/off × 2
H162 × W100 × D63
Switch rating 6A

Venner Vennerette

1	2	3	4
○	○	○	○
ON	N	L	COM
	MAINS		

Voltage free switching unless 3–4 linked

Electromechanical 24 hour time switch

On/off × 2
H102 × W87 × D76
Switch rating 30A

Venner Venomise

3	○	ON
2	○	COM
1	○	OFF
N	○	MAINS
L	○	MAINS

Electromechanical 24 hour time switch

The Venomise is designed for control of an immersion heater or similar.

On/off × 2
H113 × W109 × D50
Switch rating 16A

Venner Venneron

1	2	3	4
○	○	○	○
COM	L	N	ON
	MAINS		

Voltage free switching unless 1–2 linked

Electromechanical 24 hour time switch

On/off × 2
H115 × W88 × D82
Switch rating 15A

Venner Venotime

3	○	ON
2	○	COM
1	○	OFF
N	○	MAINS
L	○	MAINS

Voltage free switching unless L–2 linked

Electromechanical 24 hour time switch

Venotime Selective has day omission device.

On/off × 2
H113 × W109 × D50
Switch rating 16A

Venner Venotrol

Electromechanical 24 hour Basic programmer

On/off × 2
H200 × W118 × D80

Programmers and time switches

Venner Venotrol 80

A/S	5	4	3	2	1	N	L
○	○		○	○	○	○	○
ROOM STAT	CH ON		N	N	HW ON	N	MAINS

Fit link if no room stat

Electromechanical 24 hour Basic programmer

On/off × 2
H105 × W170 × D50

Venner Venotrol 80M

A/S	5	4	3	2	1	N	L
○	○	○		○	○	○	○
ROOM STAT	CH ON	CH OFF	N	HW ON	HW OFF		MAINS

Fit link if no room stat

Electromechanical 24 hour Full programmer

On/off × 2
H105 × W170 × D50

Venner Venotrol 80P

A/S	5	4	3	2	1	N	L
○	○	○		○	○	○	○
ROOM STAT	CH ON	CH OFF	N	HW ON	HW OFF		MAINS

Fit link if no room stat

Electromechanical 24 hour Full programmer

On/off × 2
H105 × W170 × D50

Venner Venotrol 80PM

A/S	5	4	3	2	1	N	L
○	○	○		○	○	○	○
ROOM STAT	CH ON	CH OFF	N	HW ON	HW OFF		MAINS

Fit link if no room stat

Electromechanical 24 hour Full programmer

On/off × 2
H105 × W170 × D50

Venner Venotrol 90

A/S	5	4	3	2	1	N	L
○	○	○	○	○	○	○	○
ROOM STAT	CH ON	CH OFF	N	HW ON	N		MAINS

Fit link if no room stat

Electromechanical 24 hour Basic programmer

On/off × 2
H105 × W170 × D50

Priority system programmers

Replacing a priority programmer with a double circuit programmer

Figure 2.12 *Horstmann 425/525 with Honeywell motorized valve and thermostats*

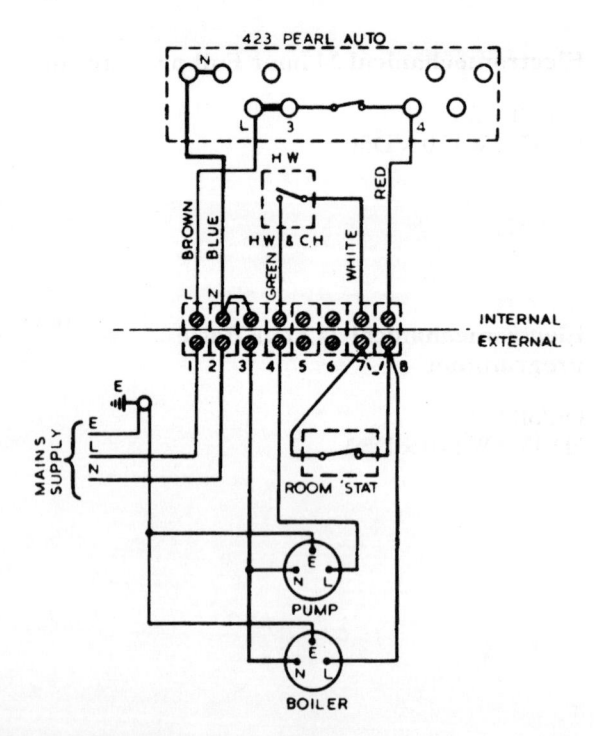

Figure 2.13 *Eberle 608*

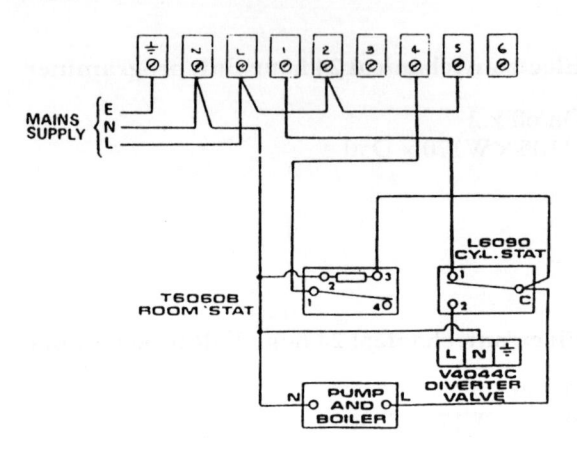

Figure 2.14 *Horstmann Sapphire 423/424*

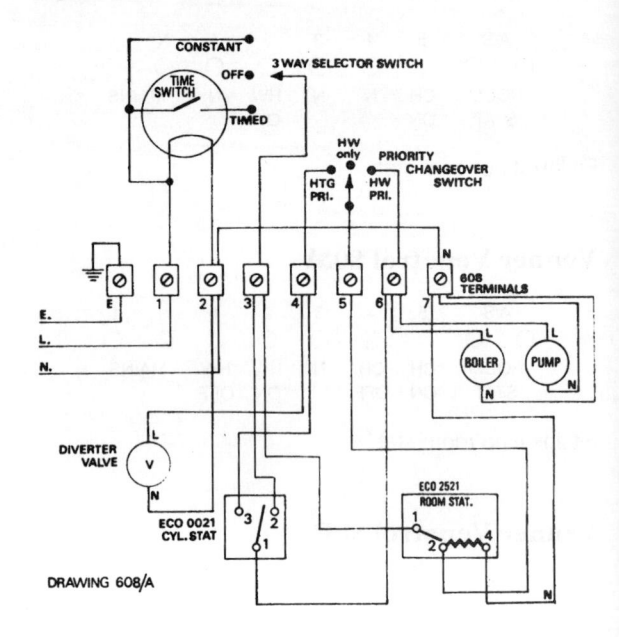

Figure 2.15 *Randall 106*

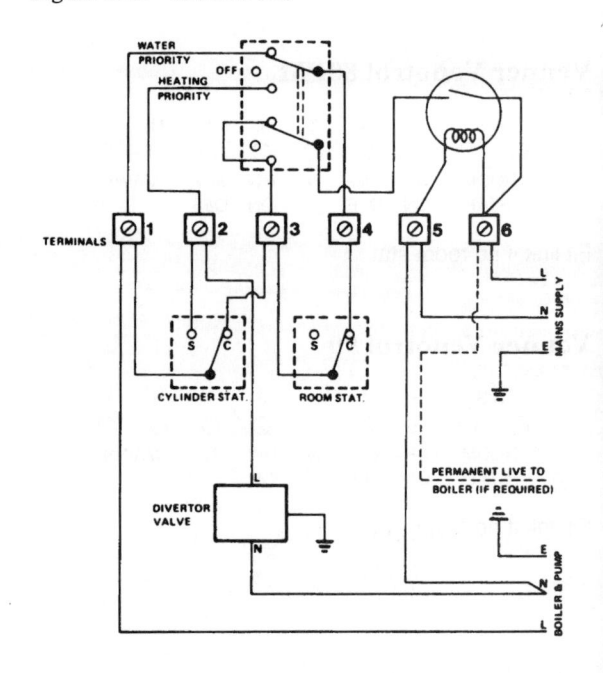

Programmers and time switches

Figure 2.16   *Randall TSR 2+2*

Figure 2.17   *Randall 3022*

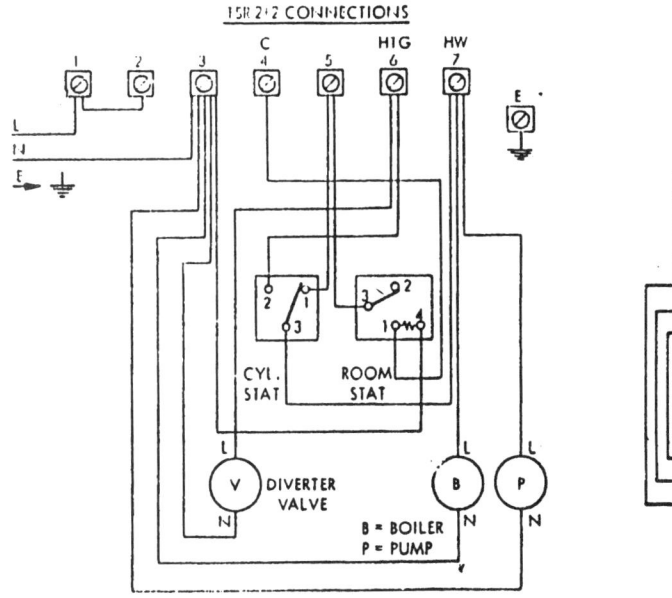

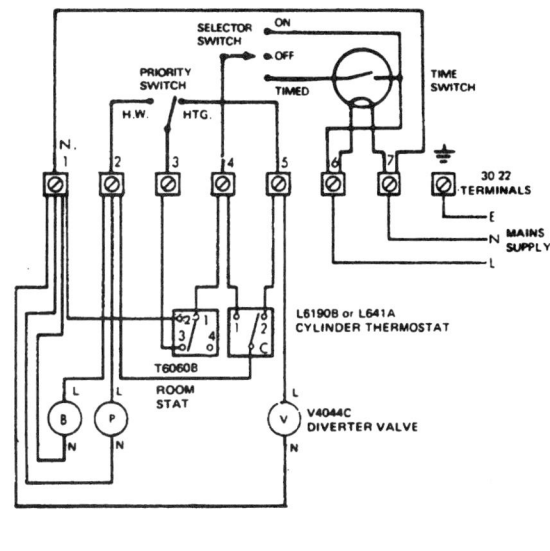

Figure 2.18   *Sangamo 409 F7*

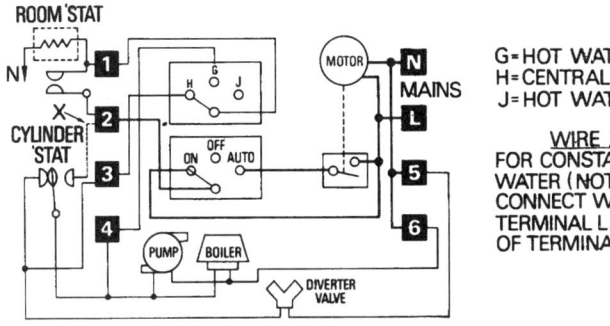

Figure 2.19   *Sangamo 410 F7*

59

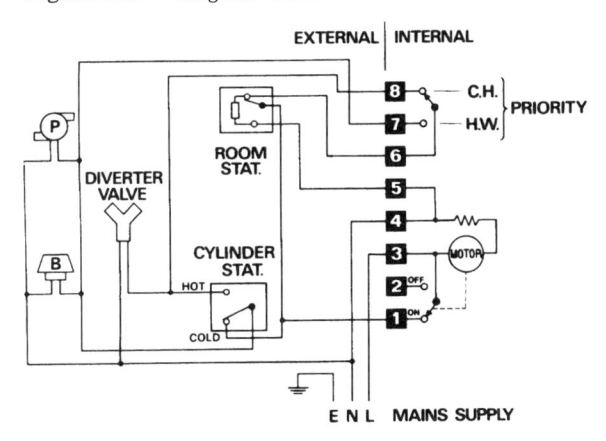

Installing and Servicing Domestic Central Heating Wiring Systems and Controls

Figure 2.20   *Smiths Centroller 30+*

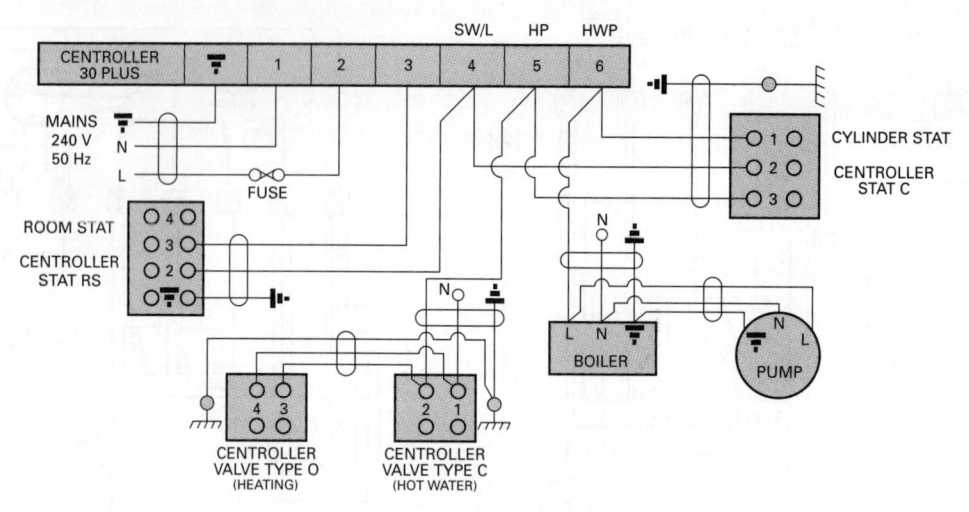

Figure 2.21   *Smiths Centroller 40+*

Figure 2.22   *Switchmaster 500*

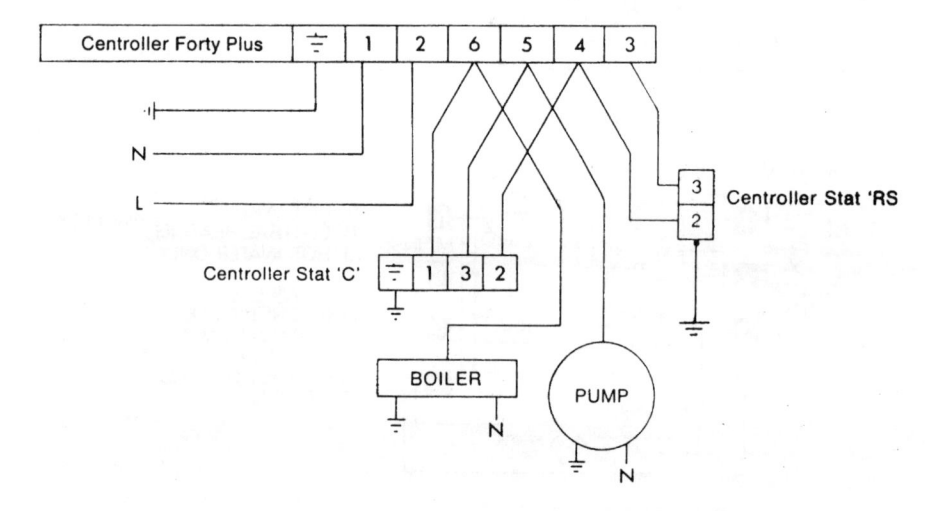

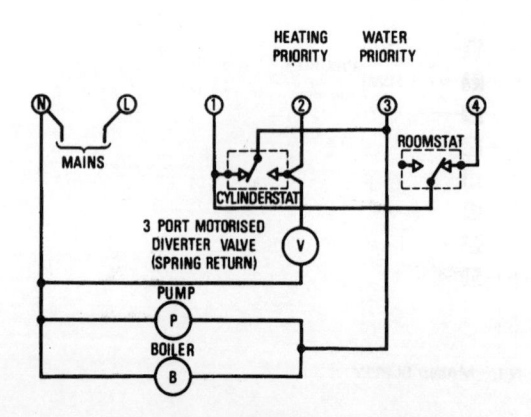

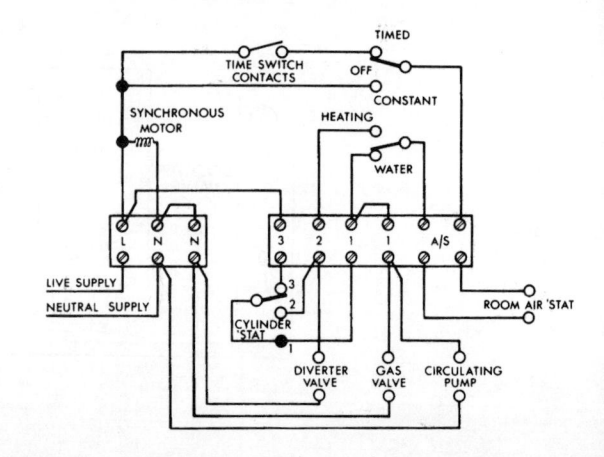

Figure 2.23   *Venner CHC/GW*

60

Programmers and time switches

Programmers specifically designed for use with the ACL 672 BR 340 Biflo motorized valve.

Others were suitable for this and other systems. See motorized valves, Chapter 6.

Figure 2.24   *Randall 105*

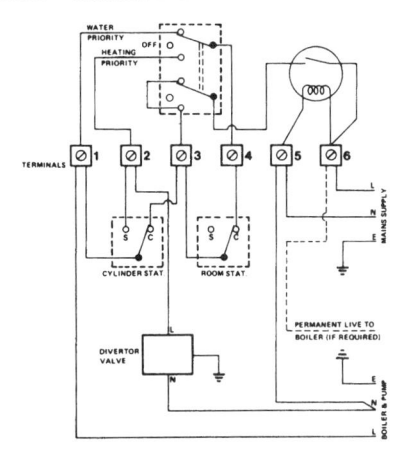

Figure 2.25   *Sangamo 410 F9*

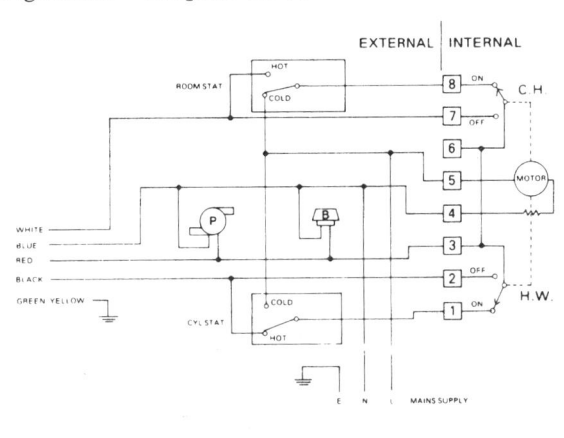

SMC Control pack wiring diagrams utilizing Horstmann programmers

Figure 2.26   *Horstmann 423 Pearl Auto*

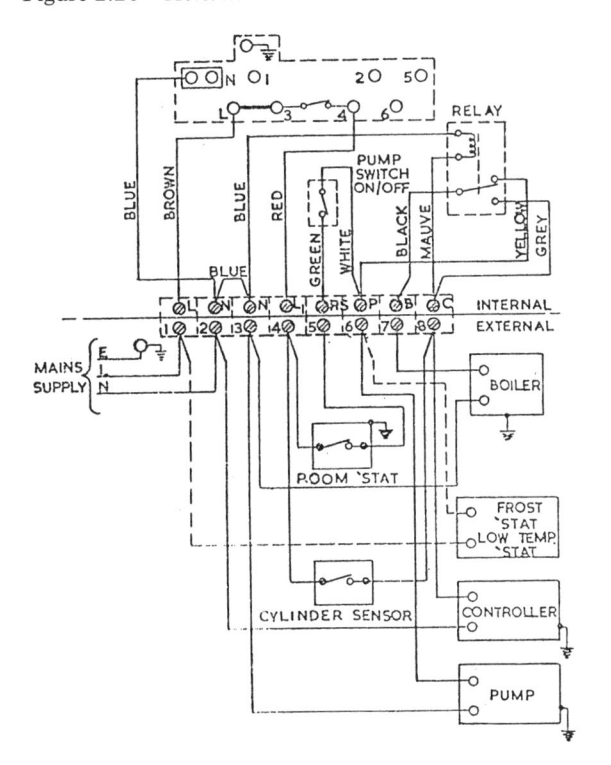

Figure 2.27   *Horstmann 424 Emerald*

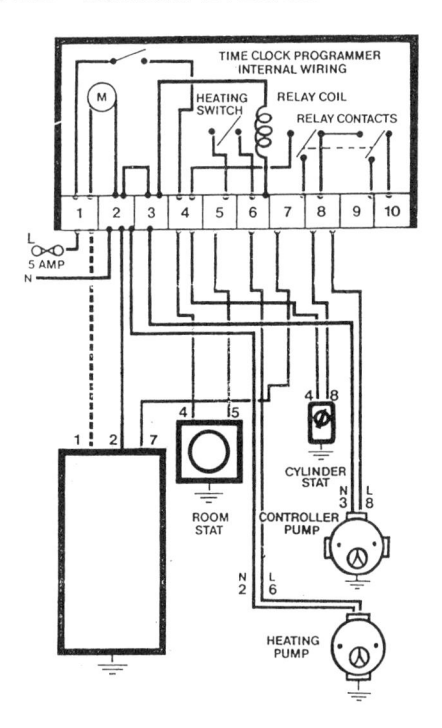

61

3

Programmers and time switches with inbuilt or external sensors or thermostats

ACL CC 240
    CC 520

Electronic comfort controller
(full control only)

The CC 240/520 allows switching times and room temperature to be selected from a central point. A remote sensor DR 174 is used to detect and control room temperature. The hot water temperature can be controlled using a conventional cylinder thermostat. The CC 240 has 24 hour programming and the CC 520 has weekday/weekend programming.

```
N   L   1      2    3    4
O   O   O      O    O    O
N   L   DR 174    HW   CH
MAINS  SENSOR    ON   ON
```

Scale 5–30°C
H93 × W148 × D31
Switch rating 2A (1A)
On/offs CC 240 × 2
        CC 520 × 3

Terminals 4 and 5 are internally linked

ACL CT 171
    CT 172

Programmable electronic thermostats with low temperature protection

The CT 171/172 allow switching times and room temperature to be selected from a single point. An inbuilt sensor is used to detect and control room temperature. In the OFF position central heating will remain off unless the temperature drops to the low temperature point, in which case it will automatically switch on again.

```
N   L   1     2    3    4
O   O   O     O    O    O
N   L   COM  OFF   ON  SPARE
MAINS
```

Scale CT 171 5–30°C
      CT 172 16–28°C
Low temperature protection
    CT 171 7°C
    CT 172 16°C
H87 × W170 × D74
Switch rating 2A (1A)
On/offs × 2
Batteries required 3 × AAA

Voltage free switching unless L–1 linked

Programmers and time switches with inbuilt or external sensors or thermostats

ACL CT 174 — Electronic clock thermostat

The CT 174 allows switching times and room temperature to be selected from a central point. A remote sensor DR 174 is used to detect and control room temperature. The CT 174 has 7 day programming.

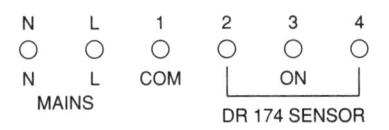

```
N    L    1    2    3    4
O    O    O    O    O    O
N    L    COM  |    ON   |
  MAINS         DR 174 SENSOR
```

Voltage free switching unless L–1 linked

Scale 16–28°C (scaled 1–5)
H87 × W170 × D47
Switch rating 2A (1A)
On/offs × 2
Batteries required 3 × AAA

ACL ILP 112 — Electronic 24 hour Basic/Full programmer

(with optimum start feature through inbuilt sensor).

```
N    L    1     2     3     4
O    O    O     O     O     O
  MAINS  HW    CH    HW    CH
         OFF   OFF   ON    ON
```

On/offs × 2
H93 × W148 × D31
Switch rating 2 (1A)
Move slider at rear of programmer to G for Basic control or P for Full control.

ACL OCC 520

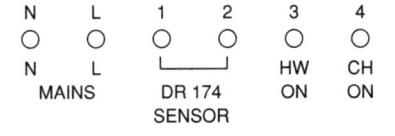

5/2 day version of OCC 720

ACL OCC 720 — Electronic 7 day Full comfort controller

(with optimum start feature)

```
N    L    1    2    3    4
O    O    O    O    O    O
N    L    |____|    HW   CH
  MAINS   DR 174    ON   ON
          SENSOR
```

On/offs × 3
H93 × W148 × D31
Switch rating 2 (1A)

ACL 0LP 552 — Electronic weekday/weekend Basic/Full programmer

(with optimum start feature)

```
N    L    1    2    3    4
O    O    O    O    O    O
N    L    |____|    ON   COM
  MAINS   DR 174
          SENSOR
```

On/offs × 2
H93 × W148 × D31
Switch rating 2 (1A)
Move slider at rear of programmer to G for Basic control and P for Full control.

ACL 0LP 722

7 day version of 0LP 522

ACL OPT 170

Electronic clock thermostat
(with optimum start feature)

The OPT 170 allows switching times and room temperatures to be selected from a central point. A remote sensor DR 174 is used to detect and control room temperature. The optimum start feature automatically reduces warm up time for the heating system as weather becomes milder.

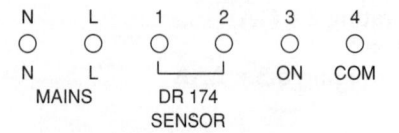

```
N   L    1    2    3    4
O   O    O    O    O    O
N   L    |____|    ON   COM
MAINS    DR 174
         SENSOR
```

Scale 5–30°C
H93 × W148 × D31
Switch rating 2A (1A)
On/offs × 2

Voltage free switching unless L–4 linked

ACL PT 110
PT 170

Programmable electronic thermostat

The PT 110/170 allows switching times and room temperature to be selected from a central point. A remote sensor DR 174 is used to detect and control room temperature. The PT 110 has a 24 hour programme and the PT 170 has a 7 day programme.

```
N   L    1    2    3    4
O   O    O    O    O    O
N   L    |____|    ON   COM
MAINS    DR 174
         SENSOR
```

Scale 5–30°C
H93 × W148 × D31
Switch rating 2A (1A)
On/offs × 2

Voltage free switching unless L–4 linked

ACL PT 271
PT 371

Programmable electronic thermostat

The PT 271/371 allows switching times and room temperatures to be selected from a central point. An inbuilt sensor is used to detect and control room temperature. PT 271 programmer options are off/time/high/low. When set to timed the programmer will turn on or off at the pre-programmed times and temperatures. There are four temperature setting periods. The PT 371 programme options are low/timed/medium/high. There are six temperatures setting periods.

```
N   L    1    2    3    4
O   O    O    O    O    O
N   L   COM  OFF   ON  SPARE
MAINS
```

Scale 5–30°C
H93 × W148 × D31
Switch rating 2A (1A)

Voltage free switching unless L–1 linked

ACL-Drayton
Digistat 2

24 hour programmable room thermostat

```
    1    2    3
    O    O    O
   COM  SAT  DEM
```

Scale 5–30°C
H87 × W87 × D33
Switch rating 2A (1A)
On/offs × 2
Batteries required 2 × AA

Programmers and time switches with inbuilt or external sensors or thermostats

**ACL-Drayton
Digistat 3**

24 hour programmable room thermostat

7 day version of Digistat 2.

Danfoss Randall WP75H

						S1	S2
○	○	○	○	○	○	○	○
E	N	N	OFF	COM	ON	Remote Sensor	

5/2 day or 7 day programmable hot water thermostat

Scale 35–65°C
H88 × W135 × D43
Switch rating 16A (4A)
On/offs × 2
Batteries required 2 × AA

Drayton Digistat RF 2

N	L	1	2	3
○	○	○	○	○
MAINS		COM	SAT	DEM

Electronic wireless 24 hour programmable thermostat

Scale 5–30°C
H87 × W87 × D33
Switch rating 2A (1A)
On/offs × 2
Batteries required 4 × AA

Drayton Digistat RF 3

Electronic wireless 24 hour programmable thermostat

7 day version of Digistat RF 2

**Eberle RTR-UTQ 9200
RTR-UWQ 9400E**

Electromechanical clock thermostat

This is a room thermostat with a built-in time switch which reduces temperature at selected times. It incorporates a three-position switch – clock, moon and sun. With the switch in the clock position, the time switch will automatically switch from set temperature to reduced temperature by 2–10°C. In the moon position the time switch will override the clock to give continuous set-back. In the sun position, the time switch will override the clock to give continuous temperature as set on the thermostat.

5	6	1	3	2	4
○	○	○	○	○	○
	L	COM	OFF (9400 ONLY)	ON	N

Scale 5–30°C
H71 × W142 × D32
Switch rating 10A (4A)
On/offs × 3

Link 1–6 if required

Installing and Servicing Domestic Heating Wiring Systems and Control

Eberle RTR-UTQ 9230   24V electromechanical clock thermostat

This is a room thermostat with a built-in time switch which reduces temperature at selected times. It incorporates a three-position switch – clock, moon and sun. With the switch in the clock position, the time switch will automatically switch from set temperature to reduced temperature by 2–10°C. In the moon position, the time switch will override the clock to give continuous set-back. In the sun position, the time switch will override the clock to give continuous temperature as set on the thermostat.

```
  5      6      1      3      2      4
  O      O      O      O      O      O
L-24V   COM   SPARE          ON    N-24V
```

Link 1–6 if required

Scale 5–30°C
H71 × W142 × D32
Switch rating 1A (1A)
On/offs × 3

Eberle RTR-UQW 9300

Electromechanical clock thermostat

7 day version of RTR-UTQ 9200 with 9 on/offs.

Honeywell CM 41

Electronic programmable thermostat

```
   A        B
   O        O
 LIVE     LOAD
  IN
```

Scale 5–30°C
H80 × W130 × D37
Switch rating 8 (3A) SPST
4 time/temperature pairs per day
Batteries required 2 × AA

Honeywell CM 51

7 day electronic programmable thermostat

```
   A        B
   O        O
 LIVE     LOAD
  IN
```

Scale 5–30°C
H80 × W130 × D37
Switch rating 8 (3A) SPST
6 time/temperature pairs per day
Batteries required 2 × AA

Honeywell CM 4000
CM 5000

Electronic programmable thermostat

The CM 4000/5000 allows switching times and room temperature to be selected from a central point. An inbuilt sensor is used to detect and control room temperature. The CM 4000 has a 24 hour programming and the CM 5000 has 7 day programming.

```
    A      B      C
    O      O      O
 L(COM)   ON    OFF
```

As the unit is battery operated no neutral is required

Scale 8–32C
H82 × W131 × D35
Switch rating 3A (3A)
Setting periods × 6
Batteries required 2 × AA

Programmers and time switches with inbuilt or external sensors or thermostats

Horstmann
Centaurstat 1
Centaurstat 7

Electronic programmable thermostat

The Centaurstat allows switching times and room temperatures to be selected from a central point. An inbuilt sensor is used to select and control room temperature. The Centaurstat 1 has 24 hour setting and the Centaurstat 7 has weekday/weekend setting.

```
4  O   SPARE
3  O   OFF
2  O   ON
1  O   ON
```

Scale 6–30°C
H71 × W142 × D30
Switch rating 8A (3A)
Setting periods × 4
Batteries required 3 × AA

As the unit is battery operated no neutral is required

Landis & Gyr RAV 1 Chronogyr

Electromechanical clock thermostat

The RAV 1 is a room temperature controller with adjustable automatic night set-back. Two knobs are provided for setting the day and night room temperatures and a sychronous time switch for adjustment of the changeover from day/night temperature times.

```
1   2   3    4   5   6
O   O   O    O   O   O
L   N   COM  ON  N   E
MAINS
```

Scale 5–35°C
H75 × W153 × D70
Switch rating 4A (2.5A)

Link 1–3 if required. Link 2–5

Landis & Gyr
Chronogyr   RAV 10U
                 RAV 10Ure 24V

Electromechanical clock thermostat

The RAV 10 is a room thermostat with a built-in time switch which reduces temperature at selected times. It incorporates a three-position switch – clock, moon and sun. With the switch in the 'clock' position, the time switch will automatically switch from set temperature to reduced temperature by 2–12°C. In the 'moon' position, the time switch will override the clock to give continuous set-back. In the 'sun' position, the time switch will override the clock to give continuous temperature as set on the thermostat.

```
1    6    7    5    2    3    4
O    O    O    O    O    O    O
L    L    N    N         OFF  ON
 LINK      LINK
```

Scale 5–30°C
H80 × W160 × D28
Switch rating 6A

Installing and Servicing Domestic Heating Wiring Systems and Control

Landis & Gyr RAV 91 Chronogyr   Electromechanical clock thermostat

The RAV 91 is a room thermostat with a built-in time switch which reduces temperature at selected times. It incorporates a three-position switch – clock, moon and sun. With the switch in the 'clock' position, the time switch will automatically switch from set temperature to reduced temperature by 2–12°C. In the 'moon' position, the time switch will override the clock to give continuous set-back. In the 'sun' position, the time switch will override the clock to give continuous temperature as set on the thermostat. The unit has a quartz clock and is powered by batteries.

```
Q2      Q1                    Scale 8–27°C
O       O                     Switch rating 6A (2.5A)
ON      COM
```

Landis & Gyr REV 10 Chronogyr   Electronic programmable room thermostat

The REV 10 is a room thermostat with a built-in time switch which reduces temperature at selected times. It incorporates a five-position switch – A, B, sun, moon and off. With the switch at A or B the function will be auto operation to heating programme A (two-set back cycles per 24 hours) or B (one set-back cycle per 24 hours). In the sun position, the timeswitch will override the clock to give continuous temperature as set on the thermostat, and the moon position will give continuous temperature at the pre-set reduced temperature. In the off position the heating will be off unless the frost protection facility, set at 5°C, takes over. The REV 10 has 24 hour setting facility.

```
Q2      Q1                    Scale 3–29°C
O       O                     H89 × W115 × D25
ON      COM                   Switch rating 10A (5A)
    24–250V ac                Setting periods up to 2 set-backs per day
                              Batteries required 2 × AA
```

Landis & Gyr REV 20 Chronogyr   Electronic programmable room thermostat

The REV 20 is a room thermostat with a built-in time switch which reduces temperature at selected times. It incorporates a five-position switch – A, B, sun, moon and off. With the switch at A or B the function will be auto operation to heating programme A (two set-back cycles per 24 hours) or B (one set-back cycle per 24 hours). In the sun position, the timeswitch will override the clock to give continuous temperature as set on the thermostat, and the moon position will give continuous temperature at the pre-set reduced temperature. In the off position the heating will be off unless the frost protection facility, set at 5°C, takes over. The REV 20 has 7 day setting facility.

```
Q1     Q2     Q3              Scale 3–29°C
O      O      O               H90 × W115 × D32
COM    ON     OFF             Switch rating 5A (2A)
     24–250V ac               On/off × 3 per day
                              Batteries required 3 × AA
```

Landis and Staefa REV 11 Chronogyr   Electronic programmable room thermostat

As REV 15 with daily operation

Programmers and time switches with inbuilt or external sensors or thermostats

Landis & Staefa REV 15/15T Chronogyr — Electronic programmable room thermostat

The REV 15 incorporates a self learning control algorithm that enables the thermostat to adjust to local climate, building and heating installation environments to provide optimum comfort benefits. The REV 15T has a remote operation facility. Both units have 5/2 day operating modes and are battery powered.

```
 L      L1
 O      O
COM    DEM
```

Scale 0–40°C
H104 × W128 × D37
Switch rating 8A (3.5A)
Batteries required 2 × AA

Landis & Staefa REV 22/22T Chronogyr — Electronic programmable room thermostat

The REV 22 incorporates a self-learning control algorithm that enables the thermostat to adjust to local climate, building and heating installation environments to provide optimum comfort benefits. The REV 22T has a remote operation facility. Both units have 7 day programming and holiday programming with three comfort settings per day. Also incoporates night set-back and frost protection. The units are battery powered.

```
 L    L1   L2
 O    O    O
COM  N.O. N.C.
```

Scale 0–40°C
H104 × W128 × D37
Switch rating 6A (2.5A)
Batteries required 2 × AA

Potterton PET 1 — Electronic programmable thermostat

The PET 1 allows switching times and room temperatures to be selected from a central point. An inbuilt sensor is used to detect and control room temperature. The PET 1 has weekday/weekend programming facility and low limit frost protection override.

```
 2      L       1      N      E
 O      O       O      O      O
OFF   L(COM)   ON      N      E
```

Scale 6–29°C
H88 × W142 × D46
Switch rating 6A (1A)
Setting periods × 4 in 24 hours
Batteries required 4 × AA

Randall TP 1

As Potterton Pet 1

Randall TP 2, TP 3, TP 4, TP 5 — Electronic programmable thermostat

The TP2–5 allows switching times and room temperatures to be selected from a central point. An inbuilt sensor is used to detect and control room temperatures. The TP 2 and TP 4 have 24 hour programming and the TP 3 and TP 5 have weekday/weekend programming.

```
 3    2    1
 O    O    O
ON   OFF  COM
```

As the unit is battery powered no neutral is required

Scale TP 2 and TP 3 16–30°C
Scale TP 4 and TP 5 5–30°C
H81 × W98 × D34
Switch rating 6A
Setting periods × 6 in 24 hours
Batteries required 2 × AA

Installing and Servicing Domestic Heating Wiring Systems and Control

Randall TP 6, TP 7

Electronic programmable thermostat with timed hot water Basic/Full control

The PT 6 and TP 7 allow switching times and room temperatures to be controlled from a central point. An inbuilt sensor is used to detect and control room temperatures. The hot water temperature can be controlled using a conventional cylinder thermostat. For Basic control, remove white socket from back of unit.

```
E   N   L   1    2    3    4   5    6
O   O   O   O    O    O    O   O    O
E   N   L   HW   HW   HW   CH  CH   CH
   MAINS    ON   COM  OFF  ON  COM  OFF
```

Scale TP 6 16–30°C
Scale TP 7 5–30°C
H105 × W150 × D38
Switch rating 3A
CH setting periods × 6 in 24 hours
HW on/offs × 2

Randall TP 8, TP 9

Electronic programmable thermostat with timed hot water Basic/Full control

The TP 8 and TP 9 allow switching times and room temperatures to be controlled from a central point. A remote sensor is used to detect and control room temperature. The hot water can be controlled using a conventional cylinder thermostat. For Basic control, remove white socket from back of unit.

```
E   N   L   1    2    3    4   5    6    7    8
O   O   O   O    O    O    O   O    O    O    O
E   N   L   HW   HW   HW   CH  CH   CH    SENSOR
   MAINS    ON   COM  OFF  ON  COM  OFF
```

Link L–2–5 if required

Scale TP 8 16–30°C
Scale TP 9 5–30°C
H105 × W150 × D38
Switch rating 3A
CH setting periods × 6 in 24 hours
HW on/offs × 2

Smiths Centroller 2000

Electronic 24 hour Basic/Full programmer

This programmer was supplied with room and cylinder sensors. The on/off times are divided up into day/nightime periods and room and water temperature are set on the programmer itself.

Two diagrams are shown; note that boiler and pump switching were done by the programmer and not the auxiliary switches of motorized valves.

H125 × W170 × D70
Switch rating 5A (2A)

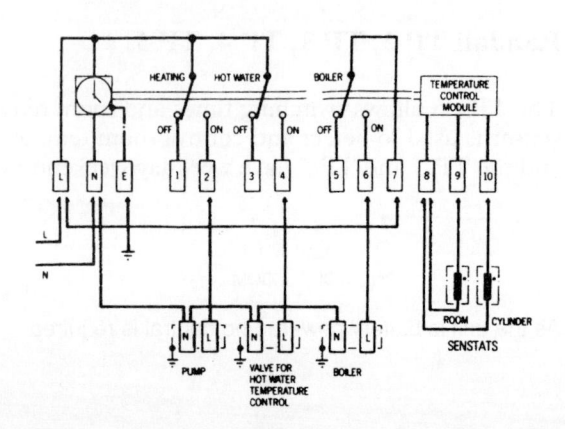

Figure 3.1 *Centroller 2000/3000. Basic system pumped heating gravity hot water. Do NOT use on low water content boilers.*

Programmers and time switches with inbuilt or external sensors or thermostats

Figure 3.2   *Centroller 2000/3000. Fully pumped 2 valve zone control. For spring return motorized valves ignore terminals 1 and 3.*

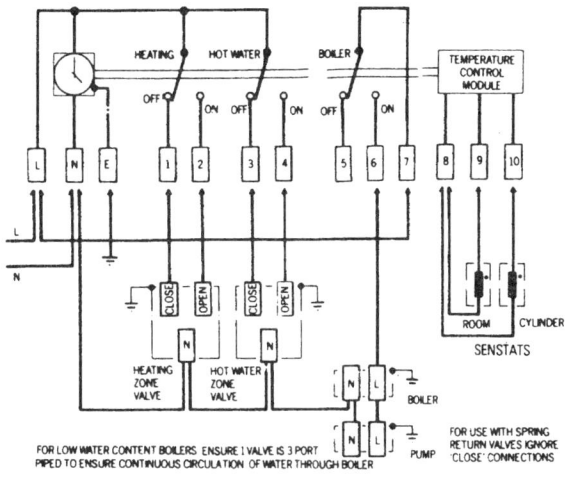

FULLY PUMPED 2 VALVE ZONE CONTROL

Smith Centroller 3000   **Electronic 24 hour Basic/Full programmer**

As Centroller 2000 with boost facility for hot water.

Smiths ERS 1   **Electronic programmable thermostat**

The ERS 1 allows switching times and room temperatures to be selected from a central point. An inbuilt sensor is used to detect and control room temperature. The ERS 1 has weekday/weekend programming facility and low limit frost protection override.

```
  2      L      1      N      E
  O      O      O      O      O
 OFF  L(COM)   ON      N      E
```

Scale 6–29°C
H88 × W142 × D46
Switch rating 6A (1A)
Setting periods × 4 in 24 hours
Batteries required 4 × AA

Sunvic EC 1401/1451   **Electronic clock thermostat**

The Sunvic EC electronic clock thermostat is designed to control room temperature at two pre-selected levels – day and night. During the on periods the temperature is controlled by the selected day temperature and at other times controlled to the night temperature setting.

```
  1    2    3    4    5    6    7    8
  O    O    O    O    O    O    O    O
  L    N   OFF  ON   COM | COM  ON  OFF
 MAINS      CIRCUIT A    |   CIRCUIT B
```

Scale 10–40°C
H110 × W180 × D65
Switch rating 5A (1A)

Voltage free switching unless L–5–6 linked

Sunvic TLC 2358 — Electromechanical clock thermostat

The TLC is a room thermostat with a built-in time switch which reduces temperature at selected times. It incorporates a three-position switch – clock, moon and sun. With the switch in the clock position, the time switch will automatically switch from set temperature to reduced temperature by 6°C. In the moon position, the time switch will override the clock to give continuous set-back. In the sun position, the time switch will override the clock to give continuous temperature as set on the thermostat.

```
   1     2      3      4
   O     O      O      O
   ON   OFF   L(COM)   N
                MAINS
```

Scale 3–27°C
H87 × W157 × D47
Switch rating 2A (1A)

Switchmaster Serenade — Electronic programmable room thermostat

The Serenade allows switching times and room temperature to be selected from a central point. An inbuilt sensor is used to detect and control room temperature. The Serenade has 7 day programming and a frost protection facility.

```
   4     5     6
   O     O     O
   ON   COM   OFF
```

As the unit is battery operated no neutral is required

Scale 5–29.5°C
H95 × W161 × D40
Switch rating 3A (2A)
Setting periods × 3
Batteries required 3 × AA

Switchmaster Symphony — Electronic 7 day Full programmer

The Symphony is supplied with room and cylinder sensors.

```
  L    N    1    3    4    6    A    B    C
  O    O    O    O    O    O    O    O    O
 MAINS     HW   HW   CH   CH       See Note
           ON  OFF   ON  OFF
```

A–C Room sensor. B–C Cylinder sensor

On/off × 3
H95 × W161 × D40
Switch rating 3A

Vaillant CRT 394 Calotrol — Electromechanical clock thermostat

The VRT 394 is a room thermostat with a built-in time switch which reduces temperature at selected times. It incorporates a three-position switch – clock, moon and sun. With the switch in the clock position, the time switch will automatically switch from set temperature to reduced temperature by 5°C. In the moon position, the timeswitch will override the clock to give continuous set-back. In the sun position, the timeswitch will override the clock to give continuous temperature as set on the thermostat. When tappets are pushed in, the clock functions at full set temperature, and when tappets remain out, the clock functions at reduced set temperature.

```
   7     2     1     4     3
   O     O     O     O     O
         L     N    ON    COM
        MAINS
```

Scale 5–30°C
H75 × W146 × D28
Switch rating 10A

Voltage free switching unless 2–3 linked

Programmers and time switches with inbuilt or external sensors or thermostats

The VRT 394 can also function in an on/off mode when it will turn off a boiler, for example, during the night time period. For this facility wire as below:

```
  7     2     1     4     3
  O     O     O     O     O
 COM    L     N     ON
       MAINS
```

Voltage free switching unless 2–7 linked

Vaillant VRT QT4 Calotrol
VRT QW4 Calotrol

Electromechanical clock thermostat

The QT has a 24 hour clock and the QW has a 7 day clock. They incorporate a three-position switch – clock, moon and sun. With the switch in the clock position, the time switch will automatically switch from set temperature to reduced temperature by 6°C. In the moon position, the time switch will override the clock to give continuous set-back. In the sun position, the time switch will override the clock to give continuous temperature as set on the thermostat. The set-back of 5°C can be altered up to 10°C by first isolating the power supply and removing the casing by releasing bottom screws and tilting case upwards. Adjust the potentiometer to required level and re-assemble case. When tappets are pushed in, the clock functions at full set temperature, and when tappets remain out, the clock functions at reduced set temperature.

```
  4     1     2     3
  O     O     O     O
 ON     N     L    COM
       MAINS
```

Scale 5–30°C
H75 × W142 × D35
Switch rating 2A

Voltage free switching unless 2–3 linked

Vaillant VRT-UT2-394 240V Calotrol
VRT-UT2-396 24V Calotrol

Electromechanical clock thermostat

The VRT-UT2-394/396 are room thermostats with a built-in time switch which reduces temperature at selected times. They incorporate a three-position switch – clock, moon and sun. With the switch in the clock position, the time switch will automatically switch from set temperature to reduced temperature by 5°C. In the moon position, the timeswitch will override the clock to give continuous set-back. In the sun position, the timeswitch will override the clock to give continuous temperature as set on the thermostat. When tappets are pushed in the clock functions at full set temperature and when tappets remain out the clock functions at reduced set temperature.

```
  7     2     1     4     3
  O     O     O     O     O
        L     N     ON   COM
       MAINS
```

Scale 5–30°C
H75 × W146 × D28
Switch rating 10A

Voltage free switching unless 2–3 linked

The VRT-UT2-394/396 can also function in an on/off mode when it will turn off a boiler, for example, during the night time period. For this facility wire as below:

```
  7     2     1     4     3
  O     O     O     O     O
 COM    L     N     ON
       MAINS
```

Voltage free switching unless 2–7 linked

4

Cylinder and pipe thermostats

The cylinder thermostat is a device for detecting and setting the temperature of water in the domestic hot water cylinder. It should be in a position so that the householder can easily make any adjustments that may be required. If possible, it is best to site it away from the flow/return pipes to avoiding detecting conducted heat. The thermostat is usually located a third of the way up the cylinder and virtually all are clamped to the cylinder with a metal band or spring wire. The exceptions to this rule are the Potterton PTT1 and PTT2. On these, the actual thermostat is fixed on a convenient wall and a pre-wired probe is attached to the cylinder. The probes are available with 2 metre or 10 metre leads. These thermostats are ideal where a cylinder is located in a loft, eaves cupboard, or similar difficult-to-get-at location.

Virtually all cylinder and pipe thermostats have SPDT switching and are suitable for all voltages up to 240V. Some are pre-wired and when using one of these for a SPST application the wire not used must be safely terminated as it will be live when the thermostat is in a satisfied state.

There are two instances where fixing the cylinder thermostat can be problem. These are the horizontally mounted cylinder and the square or oblong copper tank. To deal with the horizontal cylinder first. It is probably going to be a trial and error exercise as to location but a third of the way up as normal would be the best place to start. As for the square or oblong copper tank, one method is to solder two copper tags onto the tank about 12–18 inches apart and a third of the way up. Two pieces of flattened 15mm tube will do the job. Then the thermostat can be fixed using the spring wire method. Alternatively, the Potterton cylinder thermostat with sensor could be used with the sensor being held in position with suitable tape.

Pipe thermostats are usually cylinder thermostats with a modified base. They function just the same and are used for various reasons, including pump over-run thermostats.

ACL HTS

Common	3	Scale 50–80°C
Demand	1	H114 × W58 × D67
Satisfied	2	15A res.

ACL HTS 2

Common	Red	Scale 50–80°C
Demand	Black	H110 × W34 × D40
Satisfied	Yellow	3A res.

1.5 metre pre-wired

ACL HTS 2/S

As HTS 2 but metal strap fixing.

ACL HTS 3

Common	C	Scale 50–80°C
Demand	1	H100 × W40 × D45
Satisfied	2	3A res.

Barlo CT 1

Common	Red	Scale 50–80°C
Demand	Black	H100 × W34 × D40
Satisfied	Yellow	3A res.

Benefit

Common	1	Scale 15–90°C
Demand	2	H116 × W50 × D54
Satisfied	3	15A res. (2.5A)

Danfoss ATC

Common	C	Scale 20–90°C
Demand	NC	H100 × W60 × D57
Satisfied	NO	15A res. (2.5 ind.)

Danfoss ATP

Pipe thermostat

Common	C	Scale 30–90°C
Demand	NC	H100 × W60 × D57
Satisfied	NO	6A

Danfoss ATF

Pipe frost thermostat

Common	C	Scale 5–90°C
Demand	NC	H100 × W60 × D57
Satisfied	NO	6A

Drayton CS 1

Common	1	Scale 20–90°C
Demand	2	H92 × W60 × D59
Satisfied	3	6A res. (2A ind.)

Drayton CS 2

Common	1	Scale 20–90°C
Demand	2	H90 × W40 × D45
Satisfied	4	15A res. (2.5A)

Eberle RAR

Common	1	Scale 15–90°C
Demand	2	H116 × W50 × D54
Satisfied	3	15A res. (2.5A)

Honeywell L641A

Common	C	Scale 50–80°C
Demand	1	H79 × W44 × D44/54
Satisfied	2	4A res. (2A ind.)

Honeywell L641B

Pipe frost thermostat

Common	C	Scale 10–40°C
Demand	1	H79 × W44 × D56
Satisfied	2	4A (2A)

Honeywell L697A

Common	1	Scale 100–180°F
Demand	3	H120 × W94 × D50
Satisfied	2	20A res.

Honeywell L6090A

Common	C	Scale 30–90°C
Demand	1	H95 × W50 × D85
Satisfied	2	6A res. (4A ind.)

Honeywell L6190B

Common	C	Scale 25–95°C
Demand	1	H92 × W48 × D77
Satisfied	2	10A res. (2.5A)

Horstmann HCT – 1

Common	1	Scale 15–90°C
Demand	2	H116 × W50 × D54
Satisfied	3	15A res. (2.5A)

Landis & Gyr RAM 1

Common	1	Scale 15–90°C
Demand	2	H116 × W50 × D54
Satisfied	3	15A res. (2.5A)

Landis & Gyr RAM 21

Common	1	Scale 50–80°C
Demand	2	H141 × W50 × D42
Satisfied	3	6A res. (3.5A)

Potterton PTT 1

Common	L	Scale 45–75°C
Demand	H	H78 × W70 × D40
Satisfied	C	5A res. (2.5A)
Neutral	N	

This thermostat is fitted remote from the cylinder using a probe supplied with a 2 metre lead. A 10 metre lead is available. The thermostat is also supplied with indicator neons to show whether temperature is reached. The wiring of a neutral is essential.

Potterton PTT 2

From time control	TL	
Perm live/common	L	Scale 45–75°C
Demand	H	H78 × W70 × D40
Satisfied	C	5A res. (2.5A)
Neutral	N	

This thermostat has a boost facility and a permanent live is required. It is fitted remote from the cylinder using a probe supplied with a 2 metre lead. A 10 metre lead is available. The thermostat is also supplied with indicator neons to show whether temperature is reached. The wiring of a neutral is essential.

Potterton PTT 100

Common	TL	Scale 30–90°C
Demand	H	H110 × W52 × D70
Satisfied	C	16A (2A)

Proscon SOA

Common	3	Scale 33–83°C
Demand	1	H92 × W57 × D32
Satisfied	2	3A res. (2A ind.)

Randall CN4

Common	1	Scale 15–90°C
Demand	2	H116 × W50 × D54
Satisfied	3	15A res. (2.5A)

Smiths C

Common	1	Scale 120–180°F
Demand	2	H118 × W53 × D40
Satisfied	3	3A res.

Sopac SA 0570

Common	C	Scale 5–70°C
Demand	1	H100 × W50 × D40
Satisfied	2	16A res. (2.5A)

Sopac SA 2590

Common	C	Scale 25–90°C
Demand	1	H100 × W50 × D40
Satisfied	2	16A res. (2.5A)

Sopac SA 0701

Common	C	Scale 25–90°C
Demand	1	H101 × W38 × D38
Satisfied	2	16A res. (2.5A)

Sunvic PA 2252

Pipe thermostat

Common	3	Scale 40–90°C
Demand	1	H127 × W53 × D62
Satisfied	2	15A res.

Sunvic SA 36

Common	3	Scale 130–180°F
Demand	1	H105 × W55 × D63
Satisfied	2	15A res.

Sunvic SA 1452

Common	3	Scale 46–90°C
Demand	1	H100 × W50 × D50
Satisfied	2	6A res. (2.5A)

Sunvic SA 1453

As SA 1452 but pre-wired
Common	Brown	Scale 46–90°C
Demand	Black	H100 × W50 × D50
Satisfied	Blue	6A res. (2.5A)

Sunvic SA 1502

As SA 1452, pre-wired with four way plug for duo plug or clock box systems. The colour coding of the flex is:

Common	Blue
Demand	Yellow
Satisfied	White

Sunvic SA 1503

As SA 1502.

Sunvic SA 2401

Common	3	Scale 40–90°C
Demand	1	H109 × W53 × D59
Satisfied	2	15A res.

Sunvic SA 2451

Common	3	Scale 40–90°C
Demand	1	H109 × W53 × D59
Satisfied	2	15A res.

Sunvic SA 2501

As 2451, but pre-wired with four way plug fitted for duo plug system. Colour coding as SA 1502.

Switchmaster SCT 1

Common	1	Scale 40–80°C
Demand	2	H114 × W45 × D50
Satisfied	3	6A res. (1A)

Teddington FEA

Common	1	Scale 40–80°C
Demand	2	H114 × W45 × D50
Satisfied	3	6A res. (1A)

Tower CS

Common	Red	Scale 50–80°C
Demand	Black	H100 × W34 × D40
Satisfied	Yellow	3A res.

1.5 metre pre-wired

Trac TS 30

Common	1	Scale 15–90°C
Demand	2	H116 × W50 × D54
Satisfied	3	15A res. (2.5A)

Wickes CS

Common	1	Scale 50–80°C
Demand	2	H141 × W50 × D42
Satisfied	3	6A res. (3.5A)

5

Room, frost and low-limit thermostats

Room thermostats come in many shapes and sizes with different switching arrangements, current ratings and facilities such as neon indicators, off position, locking device, thermometer, night setback, etc. However, its function, when used with domestic central heating, is to act as a temperature operated switch to turn off the pump, close a motorized valve, etc.

As the thermostat is designed to operate on fluctuations in air temperature, it must be sited in a position where there is good air movement and circulation. This would normally be the hall or largest living room, about 5 feet (1.5m) from the floor. It is also necessary to ensure that it is influenced by a radiator not fitted with a thermostatic radiator valve, but it must not be sited immediately adjacent to it.

Some examples of where **not to** site a room thermostat are:

(a) within 6 inches of an internal corner as air circulates – a corner of a room is regarded as dead air space
(b) over or near an artificial heat source, e.g. table lamp, television
(c) in a kitchen or cupboard
(d) in a room containing an open fire, gas fire, electric fire, or similar heating appliance not influenced by the thermostat
(e) on an outside wall
(f) on an airing cupboard wall
(g) on a chimney breast that may be used
(h) in direct sunlight
(i) in a draught
(j) behind curtains.

Having selected the correct position, which is often a matter of compromise, ensure that the thermostat is wired up correctly. This is straightforward but it is essential for greater accuracy to wire the live into the correct terminal so that the heat anticipator, if fitted, functions correctly.

One other thing to remember is that on 24V room thermostats there is usually an adjustment to be made inside the thermostat so that the anticipator scale coincides with the current rating of the gas valve of the boiler or warm air heater. A 24V thermostat must not be used on a 240V supply, but a 240V room thermostat can be used on a 24V system, although it may perform poorly and such practice is not to be encouraged.

If part of the system is likely to suffer damage from freezing in cold weather, it is necessary to fit a frost or low-limit thermostat. Likewise, protection of the property itself is sometimes the requirement, e.g. timber framed houses may require a low-limit thermostat set at about 11C to avoid condensation forming. Usually, the frost or low-limit thermostats are wired to over-ride any time clock and room thermostat. Many electricians think that room, frost and low-limit are wired differently because one acts at high temperature and the others at low temperature. However, all three thermostats break on temperature rise and so are wired the same. When wiring a frost thermostat on a gravity hot-water/pumped central heating system, it is necessary to override both hot water and heating channels of a programmer, and so a double-pole thermostat, such as a Sopac TA 147, has to be used. An alternative is the conventional SPST thermostat, with a double-pole relay. Both methods are shown on page 211. Some manufacturers include a frost position on their normal room thermostats or have boxes for fitting the frost thermostat outside on a wall, such as the Sunvic BX3.

ACL DT

Common	3	Scale 5–25°C
Demand	1	H66 × W100 × D39
Satisfied	2	6A res. (2A)
Neutral	N	

Room, frost and low-limit thermostats

ACL ST

Common	3	Scale 5–25°C
Demand	1	H66 × W100 × D39
Satisfied	None	6A res. (2A)
Neutral	N	

ACL TA350

Common	1	Scale 6–30°C
Demand	3	H72 × W72 × D44
Satisfied	2	16A res. (2.5A)
Neutral	None	

ACL TS142

Common	1	Scale 3–30°C
Demand	2	H71 × W71 × D35
Satisfied	None	16A res. (4A)
Neutral	4	

ACL-Drayton RTS 1 and 2

Electronic thermostat, 240V only

Common	L	Scale 10–30°C
Demand	3	H86 × W86 × D37
Satisfied	None	2A (1A)
Neutral	N	RTS2 has LED 'ON' indicator

ACL-Drayton RTS 3

Frost thermostat, 240V only

Common	L	Scale 3–10°C
Demand	3	H86 × W86 × D37
Satisfied	None	2A (1A)
Neutral	N	

ACL-Drayton RTS 4

Electronic thermostat

Perm live	L	Scale 10–30°C
Common	1	H86 × W86 × D37
Demand	3	2A (1A)
Satisfied	2	
Neutral	N	

It is essential that a 240V supply is wired to L and N. Link L–1 if voltage free contacts are not required.

ACL-Drayton RTS 5

Optimum start and save facility.

Common	1	Scale 10–30°C
Demand	3	H86 × W86 × D37
Satisfied	None	2A (1A)
Neutral	N	

Barlo RT 1

Common	1	Scale 3–30°C
Demand	2	H71 × W71 × D35
Satisfied	None	16A res. (4A)
Neutral	4	

Benefit BRFT 10

Common	1	Scale 5–30°C
Demand	2	H68 × W90 × D40
Satisfied	3	16A res. (2.5A)
Neutral	None	

Brassware Ferroli

Common	2	Scale 5–30°C
Demand	1	H78 × W78 × D36
Satisfied	3	16A res. (2.5A)
Neutral	None	

Danfoss Randall RET 230 C, L, NA

Electronic thermostat

Common	L	Scale 5–30°C
Demand	3	H86 × W85 × D42
Satisfied	4	3A (1A)
Neutral	N	

Danfoss RMT 24–24V

Common	1	Scale 8–30°C
Demand	2	H80 × W80 × D40
Satisfied	3	10A (4)
Neutral	4	Neutral is – of circuit

Danfoss RMT 24T

With night set-back facility/thermometer
Common	1	Scale 8–30°C
Demand	2	H80 × W80 × D40
Satisfied	3	10A (4A)
Neutral	4+5	Neutral is – of circuit
NSB	6	

Danfoss RMT 230

Common	1	Scale 8–30°C
Demand	2	H80 × W80 × D40
Satisfied	3	10A (4A)
Neutral	4	

Danfoss RMT 230T

With night set-back facility/thermometer
Common	1	Scale 8–30°C
Demand	2	H80 × W80 × D40
Satisfied	3	10A (4A)
Neutral	4	
NSB	switch 5 and 6	

Danfoss RTF

Frost thermostat
Common	1	Scale – 5C fixed
Demand	2	H80 × W80 × D40
Satisfied	None	16A (2.5A)
Neutral	None	

Danfoss RT1

Digital electronic thermostat
Common	1	Scale 5–30°C
Demand	3	H81 × W98 × D34
Satisfied	2	6A SPDT
Neutral	None	Battery powered

Danfoss ST

Common	2	Scale 40 – 80F
Demand	3	H112 × W53 × D53
Satisfied	1	
Neutral	None	0–2 res

Danfoss TWE

Common	2	Scale 0–30°C
Demand	1	H82 × W82 × D33
Satisfied	None	10A res. (6A)
Neutral	4	

Danfoss TWK

Common	2	Scale 0–30°C
Demand	1	H82 × W82 × D33
Satisfied	3	10A res. (6A)
Neutral	4	

Danfoss TWL – 24V

Common	1	Scale 5–30°C
Demand	4	H82 × W82 × D33
Satisfied	None	1A res. (1A)
Neutral	None	

Danfoss TWLT – 24V

Night set-back facility
Common	C	Scale 5–30°C
Demand	4	H82 × W82 × D33
Satisfied	None	1A res. (1A)
Neutral	None	
NSB	6	

Danfoss TWP

Common	1	Scale 5–30°C
Demand	2	H82 × W82 × D33
Satisfied	3	10A res. (4A)
Neutral	4	

Danfoss TWPT

Night set-back facility
Common	1	Scale 5–30°C
Demand	2	H82 × W82 × D33
Satisfied	3	10A res. (4A)
Neutral	4	
NSB	5	

Danfoss TWR

Common   2   Scale 0–30°C
Demand   1   H82 × W82 × D33
Satisfied   None   10A res. (6A)
Neutral   4

Danfoss TWR 24 – 24V

Common   2   Scale 5–30°C
Demand   1   H82 × W82 × D33
Satisfied   None   1A res.
Neutral   None

Drayton Digistat 1

Electronic thermostat
Common   1   Scale 5–30°C
Demand   3   H88 × W88 × D33
Satisfied   2   2A (1A)
Neural   None
Batteries required 2 × AA

Drayton Digistat RF 1

Electronic wireless digital thermostat
Used in conjunction with SCR receiver
Common   1   Scale 5–30°C
Demand   3   H88 × W88 × D33
Satisfied   2   2A (1A)
Neutral   N   Batteries required 2 × AA
Live   L

Drayton Roomstat

Common   1   Scale 5–30°C
Demand   2   H81 × W81 × D31
Satisfied   3   10A res.
Neutral   4

Drayton RT°C

Common   3   Scale 2–27°C
Demand   1   H66 × W100 × D39
Satisfied   2   6A res. (2A)
Neutral   N

Drayton RTE

Common   1   Scale 5–30°C
Demand   2   H79 × W79 × D27
Satisfied   3   10A res. (4A)
Neutral   4

Drayton RTL – 24V

Common   3   Scale 13–24°C
Demand   1   H66 × W100 × D39
Satisfied   2   6A res. (2A)
Neutral   None

Drayton RTM

Common   3   Scale 13–24°C
Demand   1   H66 × W100 × D39
Satisfied   2   6A res. (2A)
Neutral   N

Eberle RTR3521

Common   1   Scale 3–30°C
Demand   2   H71 × W71 × D35
Satisfied   None   16A res. (4A)
Neutral   4

Eberle RTR6121

Common   1   Scale 5–30°C
Demand   2   H79 × W79 × D27
Satisfied   None   10A res. (4A)
Neutral   4

Ecko ET

Common   L   Scale 0–30°C
Demand   H   H78 × W133 × D41
Satisfied   C   20A res.
Neutral   N

Ekco ETS

With on/off switch
Common   L   Scale 0–30°C
Demand   H   H78 × W133 × D41
Satisfied   C   20A res.
Neutral   N

Ekco ET 16

Common	3	Scale 3–27°C
Demand	1	H90 × W86 × D46
Satisfied	None	16A res.
Neutral	4	

Ekco RS

Common	4	
Demand	3	H132 × W75 × D65
Satisfied	2	20A res.
Neutral	1	

Honeywell CT200

Electronic thermostat
With night set-back facility

Common	A	Scale 5–30°C
Demand	B	H84 × W84 × D34
Satisfied	C	8 (3A)
Batteries required 2 × AA		

Honeywell T87F 24V

2 wire control   Scale 5–30°C
                 Circular 93mm
                 1.5A

Honeywell T403A 1018

Common	3	Scale 40–80F
Demand	1	H98 × W83 × D40
Satisfied	None	2A res.
Neutral	2	

Honeywell T403A 1125

Frost thermostat
With tamper-proof cover

Common	3	Scale 30–70F
Demand	1	H98 × W83 × D40
Satisfied	None	1A res
Neutral	None	

Honeywell T403A 1141

Common	3	Scale 40–80F
Demand	1	H98 × W83 × D40
Satisfied	None	2A res.
Neutral	None	

Honeywell T406

Frost thermostat
With tamper-proof cover

Common	1	Scale 0–20°C
Demand	3	H87 × W114 × D42
Satisfied	None	20A res. (4A)
Neutral		

Honeywell T473X

Common	3	Scale 48–72F
Demand	2	H85 × W132 × D65
Satisfied	None	17.5A res. (6.5A)
Neutral	None	

Honeywell T498

Common	3	
Demand	1	H115 × W73 × D65
Satisfied	None	22A res.
Neutral	2	

Honeywell T603

Common	1	
Demand	3	
Satisfied	4	
Neutral	None	

Honeywell T803 – 24V

Common	3	Scale 42–80F
Demand	1	H95 × W78 × D35
Satisfied	None	0.18–0.8A res.
Neutral	None	

Room, frost and low-limit thermostats

Honeywell T822 – 24V

2 wire control   Scale 13–35°C
H125 × W79 × D31
0.18–0.8A res.

Honeywell T832 1083 – 24V

Incorporates two temperature setting levers and a hand wound clock, which enables user to select day and night running temperatures.

Circular in shape 93mm

Honeywell T4160A

Frost thermostat
With tamper-proof cover
Common	1	Scale 0–20°C
Demand	3	H79 × W83 × D49
Satisfied	None	2A res. (2A)
Neutral	None	

Honeywell T4160C

With night set-back facility
Common	1	Scale 10–30°C
Demand	3	H79 × W83 × D49
Satisfied	None	2A res. (2A)
Neutral	None	
NSB	5	

Honeywell T4360A

Frost thermostat
Common	1	Scale 3–22°C
Demand	3	H84 × W84 × D42
Satisfied	None	10A (3A)
Neutral	None	

Honeywell T4360B 1015

Common	1	Scale 10–30°C
Demand	3	H84 × W84 × D42
Satisfied	None	16A
Neutral	None	

Honeywell T4360E

With night set-back facility
Common	1	Scale 10–30°C
Demand	3	H84 × W84 × D42
Satisfied	None	10A res. (3A)
Neutral	2 and 5	
NSB	6	

Honeywell T6060 A/B/C

T6060A – *no anticipator*
T6060B – *with anticipator and optional thermometer*
T6060C – *with anticipator, night set-back and optional thermometer*
Common	1	Scale 10–30°C
Demand	3	H87 × W114 × D42
Satisfied	4	20A res. (4A)
Neutral	2	**(not T6060A)**

Honeywell T6160B

Common	1	Scale 10–30°C
Demand	3	H79 × W83 × D49
Satisfied	4	2A res. (2A)
Neutral	2	

Honeywell T6360B 1028

Common	1	Scale 10–30°C
Demand	3	H84 × W84 × D42
Satisfied	4	10A (3A res.) term 3
Neutral	2	6A (2A res.) term 4

Honeywell T6360B 1036

With indictor lamp illuminated when calling for heat
Common	1	Scale 10–30°C
Demand	3	H84 × W84 × D42
Satisfied	4	10A (3A res.) term 3
Neutral	2	6A (2A res.) term 4

Honeywell T6360B 1069

With tamper-proof cover
Common	1	Scale 10–30°C
Demand	3	H84 × W84 × D42
Satisfied	4	10A (3A res.) term 3
Neutral	2	6A (2A res.) term 4

Honeywell T6360B 1085

Common	1	Scale 1–5°C
Demand	3	H84 × W84 × D42
Satisfied	4	10A (3A res.) term 3
Neutral	2	6A (2A res.) term 4

Horstmann HRT 1

Common	1	Scale 3–30°C
Demand	2	H71 × W71 × D35
Satisfied	None	16A res. (4A)
Neutral	4	

KDG Range

As Sopac/Smiths Industries.
For model see inside cover.

Landis & Gyr RAD 1A

Common	1	Scale 5–30°C
Demand	2	H78 × W78 × D43
Satisfied	3	16A res. (2.5A)
Neutral	4	

Landis & Gyr RAD 1F

Common	1	Scale 5C fixed
Demand	2	H78 × W78 × D43
Satisfied	3	16A res. (2.5A)
Neutral	None	

Landis & Gyr RAD 1N

Common	1	Scale 5–30°C
Demand	2	H78 × W78 × D43
Satisfied	3	16A res. (2.5A)
Neutral	None	

Landis & Gyr RAD 5

Common	6	Scale 5–30°C
Demand	2	H80 × W80 × D27
Satisfied	None	6A res. (2.5A)
Neutral	4	

Landis & Gyr RAD 7

Common	1	Scale 5–30°C
Demand	2	H82 × W82 × D30
Satisfied	None	15A res. (4A)
Neutral	4	

Nettle

The Nettle range of thermostats is similar to the Honeywell T6060 range.

Pegler SR 2

Common	L	Scale 0–30°C
Demand	H	H78 × W133 × D41
Satisfied	C	20A res.
Neutral	N	

Potterton PRT 1

Common	L	Scale 5–30°C
Demands	H	H78 × W70 × D38
Satisfied	None	5A res. (2.5A)
Neutral	N	

This thermostat has indicators to show whether the set temperature has been reached. The wiring of a neutral is essential.

Potterton PRT 2

Common	TL4	Scale 4–30°C
Demand	H3	H84 × W110 × D38
Satisfied	None	5A res. (2.5A)
Neutral	N5	

If the thermostat is to be used for 240V, link TL to COM. If the thermostat is to be used for switching a different voltage, e.g. 24V, the switch contacts are Terminal COM and H and the above link should not be fitted. However, a 240V supply to Terminal TL must be fitted and a neutral is required in all applications. This thermostat is fitted with indicators to show whether the set temperature has been reached.

Room, frost and low-limit thermostats

Potterton PRT 100 DT

Common	TL	Scale 5–30°C
Demand	H	H70 × W70 × D31
Satisfied	C	10A (3A)
Neutral	N	

Potterton PRT 100 FR

Frost thermostat

Common	L	Scale −10–15°C
Demand	H	H70 × W70 × D31
Satisfied	C	10A (3A)
Neutral	N	

Potterton PRT 100 ST

Common	TL	Scale 5–30°C
Demand	H	H70 × W70 × D31
Satisfied	None	10A (3A)
Neutral	N	

Proscon LC

Common	3	Scale 5–25°C
Demand	2	H113 × W67 × D51
Satisfied	1	16A res.
Neutral	None	

Proscon PB°C

Common	3	Scale 2–27°C
Demand	1	H66 × W100 × D39
Satisfied	2	6A res. (2A)
Neutral	N	

Proscon PBF

As Proscon PBC but scaled in Farenheit

Proscon R 1

Common	Yes	Scale 40–80°F
Demand	Yes	H128 × W73 × D61
Satisfied	Yes	20A res.
Neutral	None	

Terminals unmarked.

Randall R504D

Common	3	Scale 5–30°C
Demand	1	H84 × W84 × D45
Satisfied	2	16A res.
Neutral	N	

Randall R504N

Common	3	Scale 1–6°C
Demand	1	H84 × W84 × D45
Satisfied	2	16A res.
Neutral	N	

Randall R505D

With night set-back facility

Common	3	Scale 5–30°C
Demand	1	H84 × W84 × D45
Satisfied	2	16A res.
Neutral	N	

Randall R505N

Common	3	Scale 1–6°C
Demand	1	H84 × W84 × D45
Satisfied	2	16A res.
Neutral	N	

Randall RD 3

Common	1	Scale 5–30°C
Demand	2	H79 × W79 × D27
Satisfied	None	10A res. (4A)
Neutral	4	

Randall RSR/L

Common	3	Scale 13–24°C
Demand	1	H66 × W100 × D39
Satisfies	2	6A res. (2A)
Neutral	None	

Installing and Servicing Domestic Heating Wiring Systems and Control

Randall RSR/M

Common	3	Sale 13–24°C
Demand	1	H66 × W100 × D39
Satisfied	2	6A res. (2A)
Neutral	N	

Sangamo 925890

Common	1	Scale 5–35°C
Demand	3	H75 × W75 × D31
Satisfied	None	10A res.
Neutral	4	

Sangamo 925895

With on/off switch

Common	1	Scale 5–35°C
Demand	3	H75 × W75 × D31
Satisfied	None	10A res.
Neutral	4	

Sauter TSH 3

Common	1	Scale 4–30°C
Demand	2	H60 × W100 × D40
Satisfied	3	15A (res. (1A)
Neutral	4	

Sauter TSH 57 (F004)

Common	2	Scale 5–30°C
Demand	3	H71 × W71 × D28
Satisfied	1	10A res. (4A)
Neutral	4	

Smiths RS

Common	3	Scale 6–28°C
Demand	2	H58 × W80 × D45
Satisfied	4	10A res.
Neutral	None	

Smiths ZV2521

Common	1	Scale 0–30°C
Demand	2	H72 × W72 × D36
Satisfied	None	16A res. (4A)
Neutral	4	

Smiths ZV2522

Common	1	Scale 0–30°C
Demand	2	H72 × W72 × D36
Satisfied	5	16A res. (4A)
Neutral	4	

Sopac TA 50 range

Common	2	Scale 45–80F
Demand	3	H55 × W105 × D50
Satisfied	4	15A res.
Neutral	1	(if fitted)

Sopac TA 80

Common	3	Scale 6–28°C
Demand	2	H58 × W80 × D45
Satisfied	None	10A res.
Neutral	None	

Sopac TA 80Y

Common	3	Scale 6–28°C
Demand	2	H58 × W80 × D45
Satisfied	4	10A res.
Neutral	None	

Sopac TA 84

Common	3	Scale 6–28°C
Demand	None	H58 × W80 × D45
Satisfied	4	10A res.
Neutral	None	

Sopac TA 84Y

Common	3	Scale 6–28°C
Demand	2	H58 × W80 × D45
Satisfied	4	10A res.
Neutral	None	

Sopac TA 340

Common	1	Scale 6–30°C
Demand	3	H72 × W72 × D44
Satisfied	None	16A res. (2.5A)
Neutral	None	

Sopac TA 350

Common	1	Scale 6–30°C
Demand	3	H72 × W72 × D44
Satisfied	2	16A res. (2.5A)
Neutral	None	

Sopac TA 520

Common	1	Scale 6–30°C
Demand	3	H79 × W75 × D46
Satisfied	None	16A res. (2.5A)
Neutral	None	

Sopac TA 521

Common	1	Scale 6–30°C
Demand	3	H79 × W75 × D46
Satisfied	None	16A res. (2.5A)
Neutral	4	

Sopac TA 530

Common	1	Scale 6–30°C
Demand	3	H79 × W75 × D46
Satisfied	2	16A res. (2.5A)
Neutral	None	

Sunfine

Common	3	Scale 2–27°C
Demand	1	H66 × W100 × D39
Satisfied	2	6A res. (2A)
Neutral	N	

Sunvic TL range

T denotes with thermometer

Sunvic TL 10–24V

Common	3	Scale 38–82F
Demand	1	H67 × W103 × D65
Satisfied	None	1A res. (1A)
Neutral	None	

Sunvic TL 11–24V

Common	3	Scale 3–27°C
Demand	1	H67 × W103 × D65
Satisfied	None	1A res. (1A)
Neutral	None	

Sunvic TL 19

Common	3	Scale 3–27°C
Demand	1	H67 × W103 × D65
Satisfied	2	1A res. (1A)
Neutral	4	

Sunvic TL 25

Common	3	Scale 38–82°F
Demand	1	H67 × W103 × D65
Satisfied	2	1A res. (1A)
Neutral	4	

Sunvic TL 35

Common	3	Scale 38–82°F
Demand	1	H67 × W103 × D65
Satisfied	None	1A res. (1A)
Neutral	4	

Sunvic TL 39

Common	3	Scale 3–27°C
Demand	1	H67 × W103 × D65
Satisfied	None	1A res. (1A)
Neutral	4	

Sunvic TLM 2253

Common	3	Scale 3–27°C
Demand	1	H90 × W86 × D46
Satisfied	None	16A res.
Neutral	4	

Sunvic TLM 2257

Frost thermostat
Common	3	Scale −15–10°C
Demand	1	H90 × W86 × D46
Satisfied	None	16A res.
Neutral	None	

Sunvic TLM 2453

With tamper-proof cover
Common	3	Scale 3–27°C
Demand	1	H90 × W86 × D46
Satisfied	None	16A res.
Neutral	4	

Sunvic TLX 2222

Common	3	Scale 1–5°C
Demand	1	H90 × W86 × D46
Satisfied	None	6A res. (2.5A)
Neutral	4	

Sunvic TLX 2251 – 24V

Common	3	Scale 3–27°C
Demand	1	H90 × W86 × D46
Satisfied	None	1A res. (1A)
Neutral	None	

Sunvic TLX 2259

Common	3	Scale 3–27°C
Demand	1	H90 × W86 × D46
Satisfied	None	6A res. (2.5A)
Neutral	4	

Sunvic TLX 2356

Common	3	Scale 3–27°C
Demand	1	H90 × W86 × D46
Satisfied	2	2A res. (1A)
Neutral	4	

Sunvic TLX 2358

With tamper-proof cover
Common	3	Scale 3–27°C
Demand	1	H90 × W86 × D46
Satisfied	2	2A res. (1A)
Neutral	4	

Sunvic TLX 2360

Frost thermostat
Common	3	Scale 0–15°C
Demand	1	H90 × W86 × D46
Satisfied	None	6A res. (2.5A)
Neutral	4	

Sunvic TM 12

Frost thermostat
Common	3	Scale 8–52°F
Demand	1	H67 × W103 × D65
Satisfied	None	1A res.
Neutral	4	

Sunvic TM 16

Common	3	Scale 38–82°F
Demand	1	H67 × W103 × D65
Satisfied	None	1A res.
Neutral	4	

Sunvic TM 56

Common	3	Scale 3–27°C
Demand	1	H67 × W103 × D65
Satisfied	None	1A res.
Neutral	4	

Switchmaster plug-in

Common	3	Scale 5–30°C
Demand	2	H85 × W85 × D45
Satisfied	1	4A res. (1A)
Neutral	5	

Room, frost and low-limit thermostats

Switchmaster SRT 1/2

Common	1	Scale 6–30°C
Demand	3	H72 × W72 × D44
Satisfied	2	16A res. (2.5A)
Neutral	None	

Switchmaster SRT 3

Specification as SRT 4, but with Homewarm cover with heating/hot water scale. Can be replaced with conventional thermostat if necessary.

Switchmaster SRT 4

Common	1	Scale 6–30°C
Demand	3	H79 × W75 × D46
Satisfied	2	16A res. (2.5A)
Neutral	None	

Teddington FEB

Common	3	Scale 5–30°C
Demand	2	H85 × W85 × D45
Satisfied	1	4A res. (1A)
Neutral	5	

Thorn Security optima

Optimum start temperature control

Switch live	S/L	Scale 5–30°C
Permanent live	P/L	H84 × W84 × D30
Demand	On	6A res. (3A)
Satisfied	Off	
Neutral	N	

Permanent live is optional and provides a one hour heating boost facility.

Tower DT

Common	3	Scale 5–25°C
Demand	1	H66 × W100 × D39
Satisfied	2	6A res. (2A)
Neutral	N	

Tower SS

Common	1	Scale 0–30°C
Demand	2	H70 × W70 × D35
Satisfied	None	16A res. (4A)
Neutral	4	

Tower ST

Common	3	Scale 5–25°C
Demand	1	H66 × W100 × D39
Satisfied	None	6A res. (2A)
Neutral	N	

Trac RS10

Common	1	Scale 5–30°C
Demand	2	H78 × W78 × D43
Satisfied	3	16A res. (2.5A)
Neutral	None	

Tristat

Produced mainly for the commercial market, it is used in conjunction with a passive infra-red sensor and enables temperature to be reduced automatically, when a room is unoccupied.

Unity

Common	L	Scale 35–80F
Demand	H	H137 × W70 × D48
Satisfied	None	15A
Neutral	N	

Vaillant VRT 378

Common	3	Scale 5–30°C
Demand	4	H64 × W110 × D24
Satisfied	None	10A res.
Neutral	5	

Vaillant VRT 9090

Common	3	Scale 5–30°C
Demand	4	H60 × W112 × D35
Satisfied	None	10A res.
Neutral	5	

Installing and Servicing Domestic Heating Wiring Systems and Control

Vokera

Common	2	Scale 5–30°C
Demand	5	H80 × W80 × D35
Satisfied	None	10A res. (2.5A)
Neutral	6	
Link	1–4	

NOTE: Terminals run 1 3 2 4 5 6

Wickes RS

Common	6	Scale 5–30°C
Demand	2	H80 × W80 × D27
Satisfied	None	6A res. (2.5A)
Neutral	4	

Worcester Digistat CD

See ACL-Drayton Digistat RF1

Wylex

Common	L-in	Scale 38–80F
Demand	L-out	H122 × W90 × D50
Satisfied	None	1A ind.
Neutral	N	

6

Motorized valves and actuators

ACL Biflo	**Mid position**
672 BRO 340	¾″ BSP
679 BRO 340	22mm
773 BRO 337	1″ BSP
	Heating port B, hot water port A

This valve requires SPDT room and cylinder thermostats. It can only utilize a simple time switch or a small group of electromechanical programmers, e.g. Sangamo F410, F9, Randall 105, Tower/ACL MP, Tower/ACL FP, and Horstmann Gem (see Chapter 2). These can now be replaced with electronic programmers of the type where links usually need to be fitted live and switch commons but these links are not required for this system (see pages 61 and 202).

ACL Lifestyle	**2 port zone**	**Spring return**
679H 308	22mm	Auxiliary switch SPST 5A
779H 335	28mm	Auxiliary switch SPDT 5A

Standard colour flex conductors.

ACL Lifestyle	**Diverter**	
679H 314	22mm	Inlet centre port
779H 336	28mm	Port A open when energized (usually central heating)

Available with auxiliary switch if required. Standard colour flex conductors.

ACL Lifestyle	**Mid position**	
679H 340	22mm	Inlet centre port, heating port A, hot water port B
779H 340	28mm	

Auxiliary switch rating 5A. Standard colour flex conductors.

ACL Motortrol	**2 port zone**	**BSP**	**Spring return**
631 B308	½″ BSP	Old	Auxiliary switch SPST 5A
691 B308	½″ BSP	New	Auxiliary switch SPST 5A

672 B308	¾" BSP	Old	Auxiliary switch SPST 5A
679 B308	¾" BSP	New	Auxiliary switch SPST 5A
773 B335	1" BSP	Old	Auxiliary switch SPST 5A
779 B335	1" BSP	New	Auxiliary switch SPDT 5A

24V, 110V and energize to close – available as special. Standard colour flex conductors.

ACL Motortrol — Diverter — BSP

691 B314	½" BSP	Inlet port C, port B open when energized
672 B314	¾" BSP	(usually central heating)
679 B314	¾" BSP	Standard colour flex conductors.
773 B336	1" BSP	

Barlo 2PV 1 — 2 port zone — Spring return

As ACL 670 H308

Barlo 3PV 1 — Mid position

As ACL 679 H340

Danfoss ABV-VMT — 2 port zone — Thermohydraulic

ABV-VMT 15/8	15mm pumped only systems
ABV-VMT 22/8	22mm pumped only systems
AVB-VMT 28/8	28mm pumped only systems
AVB-VMT 15/2	15mm gravity only systems
ABV-VMT 22/2	22mm gravity only systems
ABV-VMT 28/2	28mm gravity only systems

ABV is the actuator part of the valve and is available in both 24V and 40V. They do not have an auxiliary switch.

Danfoss ABV-VMV — Diverter — Thermohydraulic

ABV-VMV-15	½"
ABV-VMV-20	¾"
ABV-VMV-25	1"
ABV-VMV-32	1¼"
ABV-VMV-40	1½"

ABV is the actuator part of the valve and is available in both 24V and 240V. They do not have an auxiliary switch. The VMV must always be installed as a mixing valve (two inlet ports) according to the flow direction arrows cast into the valve body. The VMV closes across main ports A–AB on rising spindle travel.

Danfoss DMV-2C — 2 port zone — Spring return

22mm	Auxiliary switch SPST 3A (2A)
28mm	Auxiliary switch SPDT 3A (2A)

24V version available. Standard colour flex conductors.

Motorized valves and actuators

Danfoss DMV-21   **2 port zone**   **Spring return**

1″ BSP

Auxiliary switch SPDT 3A (2A)
Wiring as DMV-2C 28mm. Standard colour flex conductors.

Danfoss DMV-3D   **Diverter**

22mm

Inlet port AB
Port A open when energized (usually central heating)
Standard colour flex conductors.

Danfoss DMV-3M   **Mid position**

22mm

Inlet port AB. Heating port A. Hot water port B
Auxiliary switch rating 3A (2A). Standard colour flex conductors.

Danfoss HP2   **2 port zone**

As Randall HP2

Danfoss HS3   **Mid position**

As Randall HS3

Drayton TA/M2   **2 port zone actuator – motor open/close**

Fits to TA/VA range of valve bodies – see after TA/M5
Energize WHITE for clockwise rotation of actuator (Port 2 shut)
Energize BLUE for anti-clockwise rotation of actuator (Port 3 shut)
Neutral BLACK
No auxiliary switch
See also page 203

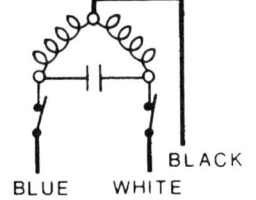

Figure 6.1   *Drayton TA/M2*

Drayton TA/M2A   **2 port zone actuator – motor open/close**

Fits to TA/VA range of valve bodies – see after TA/M5
Energise WHITE for clockwise rotation of actuator (Port 2 shut)
Energize BLUE for anti-clockwise rotation of actuator (Port 3 shut)
Neutral BLACK
Auxiliary switch SPDT 3A
Valve clockwise RED + YELLOW made
Valve anti-clockwise RED + GREY made
See also page 205

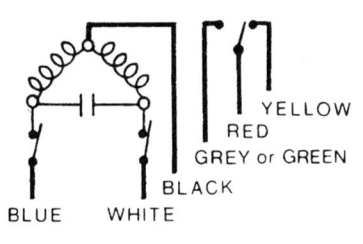

Figure 6.2   *Drayton TA/M2A*

93

Installing and Servicing Domestic Heating Wiring Systems and Control

Drayton TA/M4

Mid position actuator

Fits to TA/VA range of valve bodies – see after TA/M5.
Usually used in conjunction with an RB1 or RB2 Relay Box for boiler switching – see page 95
Energize WHITE for clockwise rotation of valve (Port 2 shut)
Energize BLUE for anti-clockwise rotation of valve (Port 3 shut)
Energize YELLOW for mid position
Neutral BLACK
See also page 204

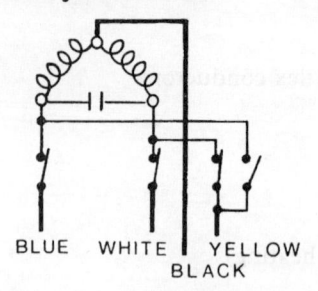

Figure 6.3   *Drayton TA/M4*

Drayton TA/M5

Diverter actuator

Fits to TA/VA range of valve bodies – see over
Energize WHITE for clockwise rotation of valve (Port 2 shut)
Energize BLUE for anti-clockwise rotation of valve (Port 3 shut)
Auxiliary switch SPDT 3A
Valve clockwise YELLOW + WHITE made
Valve anti-clockwise YELLOW + BLUE made

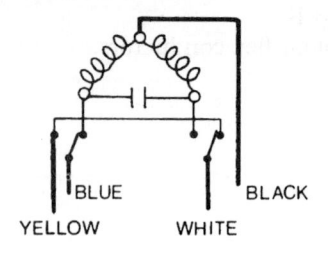

Figure 6.4   *Drayton TA/M5*

Drayton TA/VA

	Valve bodies
TA/V1	½" 15mm 2 port
TA/V2	½" 15mm 3 port inlet port 1
TA/V4*	¾" BSP 3 port, ports 2 and 3 reversible
TA/V6*	1" BSP 3 port

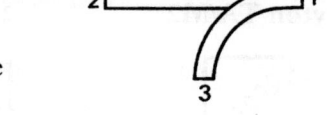

*Can be converted to 2 port by plugging third port.

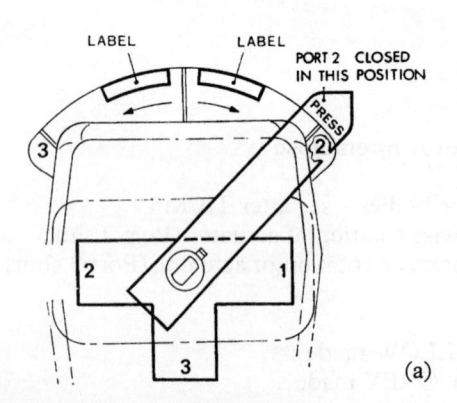

Figure 6.5   *Drayton TA/VA*

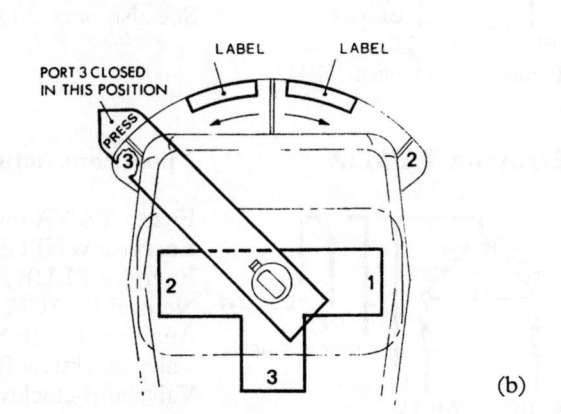

94

Motorized valves and actuators

Drayton RB1 and RB2 relay boxes (usually used with TA/M4 actuator)

Figure 6.6 *Drayton RB1*

RB1 CONFIGURATION

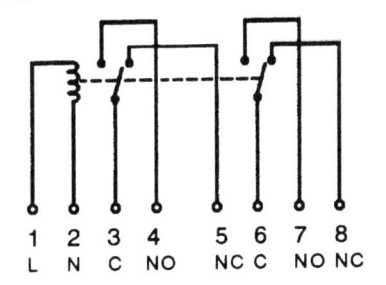

```
1  2  3  4   5  6  7  8
L  N  C  NO  NC C  NO NC
```

Double pole, double throw relay with no internal links.
H95 × W111 × D71
Contact rating 6A (1.5A)
See also page 204

Figure 6.7 *Drayton RB2*

RB2 CONFIGURATION

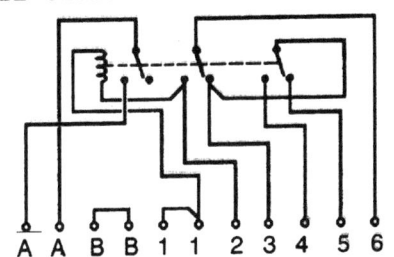

```
A A B B 1 1 2 3 4 5 6
```

Double pole, double throw relay with printed circuit internal links.

H95 × W111 × D71
Contact rating 6A (2A)
See also page 204

Drayton ZVA 22   2 port zone   Spring return

22mm   Auxiliary switch SPST 3A
                            Standard colour flex conductors

Drayton ZVA 28   2 port zone   Spring return

28mm   Auxiliary switch SPDT 3A
                            Standard colour flex conductors

Honeywell V 2057A   24V version of V 6057A

Honeywell V 4043   2 port zone   Spring return

B1257	22mm	De-energized OPEN	Auxiliary switch none – see page 197
B1265	28mm	De-energized OPEN	Auxiliary switch none – see page 197
C1156	½″ BSP	De-energized SHUT	Auxiliary switch none
H1056	22mm	De-energized SHUT	Auxiliary switch SPST 2.2A
H1106	28mm	De-energized SHUT	Auxiliary switch SPDT 2.2A
H1007	¾″ BSP	De-energized SHUT	Auxiliary switch SPST 2.2A
H1080	1″ BSP	De-energized SHUT	Auxiliary switch SPDT 2.2A

Standard colour flex conductors

Installing and Servicing Domestic Heating Wiring Systems and Control

Honeywell	V 4044C	Diverter
1288	22mm	Inlet Port AB
1569	28mm	
1098	¾" BSP	Port A open when energized
1494	1" BSP	(usually central heating)
	Standard colour flex conductors	

Honeywell V 4073A	Mid position 5 wire	
1039	22mm	Inlet port AB, heating port A, hot water port B
1088	28mm	
1054	¾" BSP	Auxiliary switch rating 2.2A
1062	1" BSP	Standard colour flex conductors

Honeywell V 4073   **Mid position 6 wire**

Inlet port Ab
Heating port A
Hot water port B

Identifiable by the external relay fitted to the valve motor cover. If replacing with V 3073 5-wire connect colour for colour (disregard brown wire connection) and change over wires on cylinder thermostat common and demand. See also pages 206 and 207

Honeywell V 6057A   **2 port zone**   **Motor open/close**

Fitted to a V 5057A valve body, ¾" or 1"
Fitted with or without SPDT
Auxiliary switch 5A (2A)
Terminal 1 Neutral
Terminal 2 Close valve
Terminal 3 Open valve

Auxiliary switch contacts if fitted:
Terminal 4   Common
Terminal 5   Made when valve shut
Terminal 6   Made when valve open

Honeywell V 8043	2 port zone
V 8044	Diverter
V 8073	Mid position

The above valves were 24 volt versions of the 240 volt range. Should any of the above need replacing, then it can be done so with the 240 V equivalent and a 24 volt motor which is available as a spare.

Hortsmann	2 port zone	Spring return
H-2-22-Z	22mm	Auxiliary switch SPST
H-2-28-Z	28mm	Auxiliary switch SPDT
	Standard colour flex conductors.	

Hortsmann	Mid-position
H-3-22-M	22mm
	Standard colour flex conductors.

Landis & Gyr Da-V322   Diverter

22mm

Inlet port AB. Port A open when energized (usually central heating)
Standard colour flex conductors.

Landis & Gyr MA-V3   Mid position

MA-V322   22mm
MA-V328   28mm

Inlet port AB, heating port A, hot water port B
Auxiliary switch rating 3A.
Standard colour flex conductors.

Landis & Gyr SK2/LL   2 port zone   Spring return

LL4402   15mm valve body
LL4453   22mm valve body
LL4501   28mm valve body
SK2   Actuator

Auxiliary switch SPST (5A(5). Standard colour flex conductors.

Landis & Gyr SK3/2701   Mid position

LT2701   22mm valve body   Inlet port AB, heating port A
SK3   Actuator   Hot water port B

Auxiliary switch rating 5A(5).   Standard colour flex conductors.

Landis & Gyr STC 4 × 7¾″   Mid position

This valve was used only in conjunction with a relay box. See page 207

Landis & Gyr ZA V2   2 port zone   Spring return

ZA-V215   15mm   Auxiliary switch SPST 3A
ZA-V222   22mm   Auxiliary switch SPST 3A
ZA-V228   28mm   Auxiliary switch SPST 3A
                   Standard colour flex conductors

Myson MSV222   2 port zone   Spring return

22mm

Auxiliary switch rating 3A
Standard colour flex conductors.

Myson MSV228   2 port zone   Spring return

28mm

Auxiliary switch rating 3A
Standard colour flex conductors.

Myson MSV322	**Mid position**	
22mm	Auxiliary switch rating 3A Inlet port AB, heating port A, hot water port B Standard colour flex conductors.	
Potterton MSV322	**Mid position**	
28mm	Auxiliary switch rating 3A Inlet port AB, heating port A, hot water port B Standard colour flex conductors.	
Potterton PMV2	**2 port zone**	**Spring return**
22mm	As Landis & Gyr SK2	
Potterton PMV3	**Mid position**	
22mm	As Landis & Gyr SK3	
Randall HP	**2 port zone**	**Spring return**
As Switchmaster VM4 auto zone valve		
Randall HP 2	**2 port zone**	**Spring return**
HPA 2	**Actuator**	Auxiliary switch SPST
HPA 2C	**Actuator**	Auxiliary switch SPDT
	Standard colour flex conductors.	
HS 3	**Diverter**	
HSA 3D	**Actuator**	
	Standard colour flex conductors.	
Randall HS	**Mid position**	
As Switchmaster VM1 Midi		
Randall HS 3	**Mid position**	
HSA 3	**Actuator** Standard colour flex conductors.	

Motorized valves and actuators

Smiths Centroller O-C/DV 1-¾"   Diverter   Thermohydraulic

This diverter valve is made up of two separate Actuators, one code O (energized to open) and one code C (energized to close). Actuator O should be in the central heating branch of the valve body.

Figure 6.8

Smiths Centroller OS/V22   2 port zone   Thermohydraulic

Consisting of a V22 or V¾" valve body and an OS energized to open actuator with SPST auxiliary switch. If no auxiliary switch required, then a Code O actuator can be used.

Figure 6.9

Smiths Centroller OS/DV 1-¾"   Mid position   Thermohydraulic

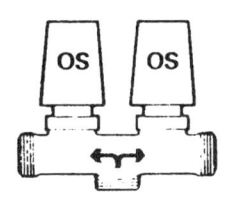

Consisting of a DV 1-¾" valve body and two OS energized to open actuators.

Figure 6.10

Smiths Centroller O-C-OS   Actuators   Thermohydraulic

Code O energize to open
Code C energize to close
Code OS energize to open with auxiliary switch

These actuators work by a heated wax cylinder, operating a plunger to open or close the appropriate port and operate an end switch if fitted. They are, therefore, wired as a spring return type actuator, but take considerably longer to operate. They were not pre-wired.

99

Sopac ZV	**2 port zone**	**Spring return**
ZV-15-2	½″ BSP	Auxiliary switch to all models
ZV-20-2	¾″ BSP	SPST switch rating 3A
ZV-20-2-B	22mm	Energize to close available with or
ZV-25-2-B	28mm	without auxiliary switches.

Sopac ZV		**Diverter**
ZV-20	¾″ BSP	Inlet port AB
ZV-25	1″ BSP	Port A open when energized (usually
ZV-20-EB	22mm	central heating).
ZV-25-B	28mm	

Sopac ZV		**Mid position**
ZV-20-EB-MID	22mm	Inlet port AB, heating port A,
ZV-25-B-MID	28mm	hot water port B
		Auxiliary switch rating 3A
		Standard colour flex conductors.

Sunvic DM 3601 **Mid position**
       DM 3551

For use only with the Duoval RJ relay box.
DM3651 has a 5-way plug for connection into the Duo-plug RJ relay box.

Sunvic DM 4601 **Mid position**
       DM 4651

For use only with the Duoval RJ relay box.
DM4651 has a 5-way plug for connection into the Duo-plug RJ relay box.
The DM 36 and 46 Duoval and Duo-plug models are interchangeable and differ only in styling. They can also be connected into any of the three relay boxes, RJ 1801, RJ 2801 and RJ 2802 or the Duo-plug relay box RJ 2852 using the appropriate wiring diagram. See pages 209 and 210

Sunvic DT and EDT	**Valve bodies for use with SD actuators**
DT 1601	¾″ BSP
DT 1701	22mm
DT 1801	Has a 28mm connection to inlet and port A and a 22mm connection to port B
DT 2601	¾″ BSP
EDT 1702	22mm
EDT 2702	22mm

Motorized valves and actuators

Sunvic Duoflow – RJ 1801 relay box

1	2	3	4	5	6	7	8	9	10	11	12	13	14	15	16
○	○	○	○	○	○	○	○	○	○	○	○	○	○	○	○
L	N	E	L	N	YEL	ORA	WHI	BLU	N	E	DEM	COM	DEM	SAT	COM
MAINS			BOILER PUMP			ACTUATOR					ROOM STAT		CYLINDER		STAT

If boiler has pump over-run, pump should be wired as boiler instructions. See also page 209

Sunvic Duoflow – RJ 2801 relay box

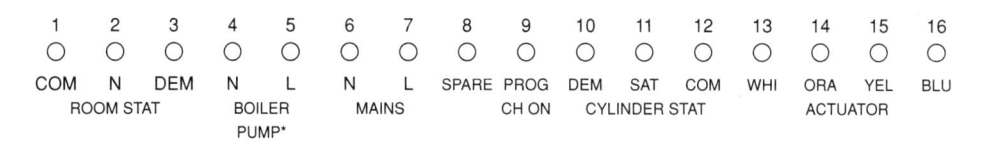

1	2	3	4	5	6	7	8	9	10	11	12	13	14	15	16
○	○	○	○	○	○	○	○	○	○	○	○	○	○	○	○
COM	N	DEM	N	L	N	L	SPARE	PROG CH ON	DEM	SAT	COM	WHI	ORA	YEL	BLU
ROOM STAT			BOILER PUMP*		MAINS				CYLINDER STAT				ACTUATOR		

If programmer has HW OFF connection wire to terminal 11.
*If boiler has pump over-run, pump should be wired as boiler instructions.
See also page 209

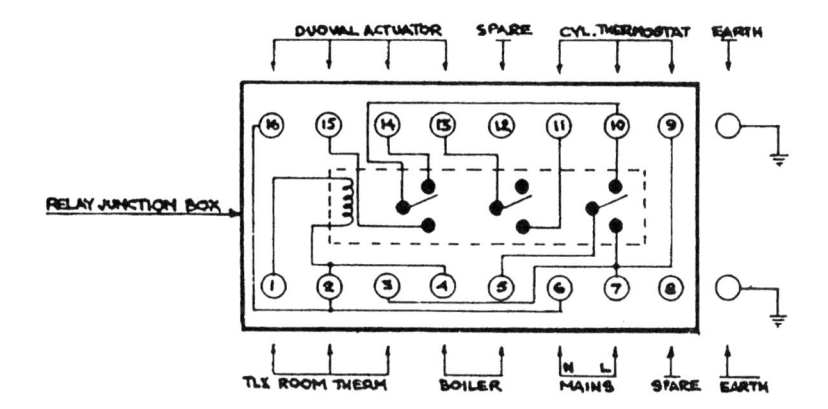

Figure 6.11 *Sunvic Duoflow RJ 2801 relay box, internal diagram.*

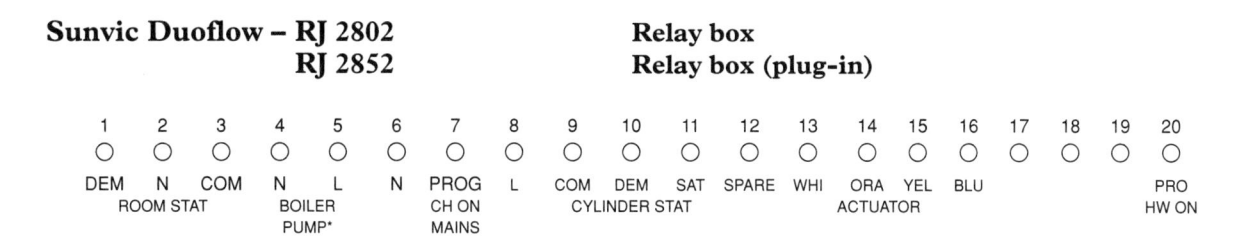

Sunvic Duoflow – RJ 2802   Relay box
RJ 2852   Relay box (plug-in)

1	2	3	4	5	6	7	8	9	10	11	12	13	14	15	16	17	18	19	20
○	○	○	○	○	○	○	○	○	○	○	○	○	○	○	○	○	○	○	○
DEM	N	COM	N	L	N	PROG CH ON MAINS	L	COM	DEM	SAT	SPARE	WHI	ORA	YEL	BLU				PRO HW ON
ROOM STAT			BOILER PUMP*						CYLINDER STAT				ACTUATOR						

If programmer has HW OFF connection wire to terminal 18.
*If boiler has pump over-run, pump should be wired as boiler instructions.
See also page 210

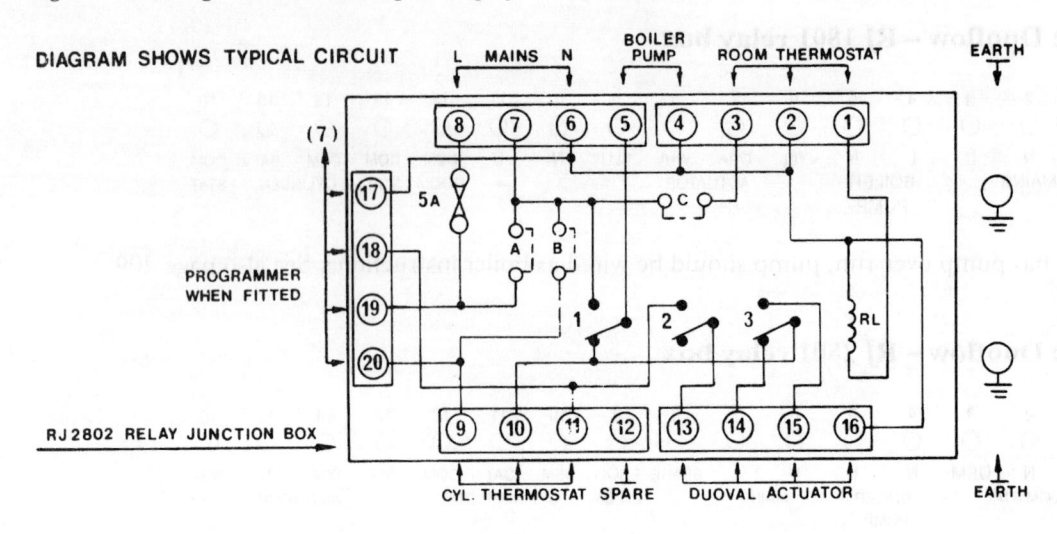

Figure 6.12 *Sunvic Duoflow RJ 2802/2852, relay box, internal diagram*

Sunvic SD 1601   **Diverter**
       SD 1626

Combines with the DT or EDT valve body to make a priority valve in the Uniflow system. The SD 1626 differs in that it has a manual lever.

Sunvic SD 1701   **Mid position**
       SD 1726

Combines with the DT or EDT valve body to make a mid position valve in the Unishare system. The SD 1726 differs in that it has a manual lever.
Standard flex covers.

Complete Valve **SDV 1211**   22mm

Sunvic SD 1752   **Mid position**

Combines with the EDT 1702 valve body to make a mid position valve in the Clockbox II Unishare system. The actuator is fitted with a four-way plug.

Sunvic 2601   Replacement actuator for SD 1601.

Sunvic SD 2701   Replacement actuator for SD 1701.

          Complete valve **SDV 2211**   22mm

Sunvic SM2, SM2201   **2 port actuator now SM 4201**

Motorized valves and actuators

Sunvic SM5, SM 2203   2 port actuator now SM 4203

Sunvic SML   2 port zone   Motor open/close

A range of 2-port motorized valves comprising of a valve body type ML and an actuator type SM to make the Minival series. The actuator range was initially SM 3201–5 and then restyled to become SM 4201/3/5 and later just SM 5201/3. The wiring principle has remained the same, although the colours were altered slightly when blue instead of black became neutral. All models fit the ML valve body.

ML 3401	½″ BSP	Valve body
ML3402	15mm	Valve body
ML3451	¾″ BSP	Valve body
ML3453	22mm	Valve body
ML3501	1″ BSP	Valve body
EML 3501	28mm	Valve body
SM 3201/4201/5201	Auxiliary switch None	18″ lead
SM 3202	Auxiliary switch None	72″ lead
SM 3203/4203	Auxiliary switch Yes	36″ lead
SM 3204/5203	Auxiliary switch Yes	17″ lead
SM 3205/4205	Auxiliary switch Yes	18″ lead

Auxiliary switch rating 3A (1.3A)

On the SM 3203/4203 and 3204 the auxiliary switch provides a live supply on the orange wire when the valve is open.
   On the SM 3205/4205 the auxiliary switch provides a live supply on the orange wire when the valve is open and a live supply on the pink wire when the valve is closed.
   The EML valve body differs from the ML range in having external compression fittings.

Colour code		*Old*	*New*
Neutral		Black	Blue
Motor open		Yellow	Yellow
Motor close		Blue	White
Aux. sw. (if fitted)	open	Orange	Orange
	closed	Pink	Pink

Sunvic SZ   2 port zone   Spring return

A range of spring return actuators that fit to the ML and EML range of valve bodies, as above, to make the Unival series.

MK1   MK2

SZ 1201/2201	Auxiliary switch None	De-energized SHUT
SZ 1226/2226	As SZ 1201, with manual lever	
SZ 151/2251	Auxiliary switch None	De-energized OPEN
SZ 1301/2301	Auxiliary switch SPST	De-energized SHUT
SZ 1302/2302	Auxiliary switch Yes	De-energized SHUT

Designed for use on gravity hot water systems. See also page 191

SZ 1326/2326	As SZ 1301, with manual lever
SZ 1327/2327	As SZ 1302, with manual lever
SZ 1351/2351	Auxiliary switch SPST De-energized OPEN
SZ 2301 F	Functionally the same as the SZ 2301 but has a superior quality motor for use in high ambient temperature locations such as within boiler casings.
SZV 2212	Complete unit
SZV 2218	Complete unit
SZV 2228	Complete unit

Switchmaster VM1   Mid position

Consisting of type VB1 valve body (22mm) and a type VA1 actuator. Known as a MIDI valve, it is blue in colour. Actuator can be turned with a screwdriver for manual operation if necessary. See also page 211

Inlet – centre T
Central heating port – right-hand with T at bottom
Hot water port – left hand with T at bottom
Auxiliary switch rating 3A

Switchmaster VM2   2 port zone   Motor open/close

Consisting of type VB2 valve body (22mm) and a type VA2 actuator, which is blue in colour. Actuator can be turned with a screwdriver for manual operation if necessary.

Auxiliary switch rating 3A

Switchmaster VM3   Mid position

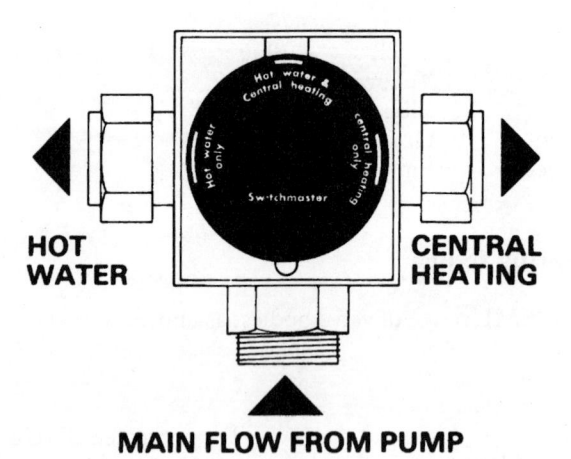

This is a manual version of the VM1 on the VB1 valve body.

Figure 6.13

Motorized valves and actuators

Switchmaster VM4   2 port zone   See description

Consisting of type VB4 valve body (22mm) and a type VA4 actuator, which is brown in colour and labelled AUTOZONE. Actuator can be turned with a screwdriver for manual operation of necessary. Although the actuator is of the motor open/close type, it is wired as and has the same colour coding as a spring return actuator, although it is necessary to ensure that the grey of the auxiliary switch is connected to live, and the orange is connected to boiler and pump.

Switchmaster VM5   Diverter

Consisting of type VB5 valve body (22mm) and a type VS5 actuator, which is brown in colour. The valve is designed specifically for the HOMEWARM system and is wired internally to give hot water priority when the room thermostat is satisfied. When the room thermostat is calling for heat, then 90% of boiler output goes to the heating circuit and the remaining 10% to the hot water circuit. A 1 metre lead is supplied, exposing four colours. However, if the cable is shortened, an orange wire will be exposed and can be disregarded.
    Same port format as VM1. See also page 206

Tower MP 3   Mid position

MP 322C   22mm
MP 3-1"B   1" BSP available special order only
MP 3-28C   28mm available special order only

Auxiliary switch rating 3A. Standard colour flex conductors. Inlet port AB, heating port A, hot water port B

Tower MV 2   2 port zone   Spring return

MV2-22C   22mm   Auxiliary switch SPST 3A
MV2-1"B   1"   Auxiliary switch SPDT 3A
MV2-28C   28mm   Auxiliary switch SPST 3A
            Standard colour flex conductors.

Tower MV3-22C   Diverter

22mm   Inlet port AB, heating port A, hot water port B
        Standard colour flex conductors.

Tower/ACL

Many of the earlier Motortrol range of motorized valves were under the TOWER and ACL name. Refer to ACL.

Wickes   2 port zone

As Landis & Gyr SK 2

Wickes   Mid position

As Landis & Gyr SK 3

7
Boilers – general

Electricity

The use of electricity as a fuel for wet system central heating systems is limited to only a few manufacturers, examples being GEC who trade under the APECS label and Redring.

Redring manufacture a range of Dualheat combination boilers for either vented or unvented applications for use in new or existing systems. The boiler provides instant hot water at all times and heating control is maintained by the use of conventional controls. It is available in 9kW or 12kW versions and uses immersion heaters to heat the water.

The GEC Nightstor central heating boilers can also operate in new or existing water radiator heating systems. They use cheap night tariff electricity to heat a solid storage core. The stored heat is extracted and fed to a conventional hot water radiator system, on demand, via the incorporated control panel and room thermostat if fitted. The hot water cylinder is heated by other means, e.g. off-peak immersion heater.

One electric boiler no longer available is the Maxton MB range manufactured by Myson Combustion Products. It was available in 6, 9 or 12kW versions and only measured 8″ × 8″ × 42″. It used immersion heaters to heat the water and was able to be used with conventional central heating controls. It relies on relays to switch the heaters on and those are prone to failure, although replacements are readily available from electrical wholesalers.

Electricty as a fuel gives the advantage of being able to site the boiler virtually anywhere as no flue or storage of fuel is required. Also they are able to utilize off-peak electricity. They do however require a substantial electrical supply, e.g. 50A in the case of the 12kW versions, and this may be a governing factor when considering options.

Gas

Gas boilers come in all shapes, sizes and formats, and as such have the advantage of being able to be utilized for virtually any application. They can be floor, wall or hearth mounted and heat exchanger material could be cast iron, copper, aluminium, stainless steel, or even a mixture of metals. Even when there is no mains gas supply, there is usually an LPG model which could be used, although the siting of the LPG storage vessel could be problematical.

Gas boilers can be used in any wet heating system and, of course, gas fuels many warm air systems that may incorporate domestic hot water circulators. In addition to a basic boiler, other models available include: boilers with integral sealed system equipment such as pressure vessel, gauge, etc., combination boilers which offer instant hot water and remove the need for a hot water storage cylinder, condensing bodies which recycle the hot flue gases to produce over 90 per cent efficiency, and System boilers which include pump, motorized valve and possibly programmer within its casing, thereby usually making installation easier.

As regards electrical control wiring, basic boilers comes in two types – those requiring just a switched live, neutral and earth connections often via an integral 3-pin plug, or the type of boiler which has a pump over-run facility. This device is a pre-set thermostat which measures the temperature of the heat exchanger. When in operation the over-run facility enables the system pump to continue running for a short period, possibly up to 20 minutes after the boiler itself has shut down thus dissipating the residual heat from the heat exchanger. This method is generally employed on wall–hung, low-water content boilers which use non cast iron heat exchangers, although some manufacturers of boilers using cast iron heat exchangers also include this over-run facility.

In addition to the above, some boilers may require additional wiring if they incorporate a programmer or timer, or in the case of back boiler units, a permanent live supply may be needed so that the live fuel effect light bulbs can be utilized when the boiler is off. It is essential that this live supply is taken from the same source as that for the boiler itself.

One type of gas boiler not mentioned in this book are those that use a pressure jet gas burner. This is because it is used on boilers intended for commercial and industrial premises. However, they can be linked into the external controls described here, in the same way as those boilers included.

Oil

Like gas, oil boilers come in a wide range of types suitable for floor, wall and hearth mounting. Oil also offers the basic boiler, combination and systems boilers. There are two types of oil boiler – wall flame and pressure jet. Wall flame boilers are the less popular and operate by oil being pumped into a large drum where it is electrically ignited, thereby heating the heat exchanger. The pressure jet fires oil into the burner under pressure and is also ignited electrically. These are the most popular type of oil boiler. Both types employ safety lock-out features to prevent oil being discharged without firing. Oil boilers need to be finely tuned in much the same way as a car engine so that it operates reliably and with maximum efficiency. It is essential that the correct grade of oil is used for each type of burner.

The method of wiring to oil boilers varies in that the supply wiring could go into the wall flame control box, the pressure jet control box or into a boiler terminal strip probably located at the top of the casing near the boiler thermostat. Wiring details are given for popular models but don't be surprised if some look as if they are not included. This may be because they employ a control box as listed and wiring details are given for those instead. Manufacturers and models where this may be the case include: Delheat, Heatrae Aggressor, McFarlane, Perrymatic and Potterton.

Solid fuel

By their very nature solid fuel boilers only come as floor standing models, although electrical controls can be incorporated into boilers fitted into fireplaces as room heaters. These latter models can have a heat exchanger for supplying hot water to the domestic cylinder or heating circuit using gravity pumped or circulation. The Honeywell Y605B Link Fuel Plan described in the ancillary equipment section (Chapter 10) may be of use to anyone wishing to install or adapt such a system. Features of a conventional solid fuel boiler may include a fan to aid combustion, or a safety switch which will shut down the boiler in the event of chimney blockage.

Because the temperature of the fuel cannot be instantly reduced or shut down as with gas or oil, solid fuel boilers cannot be employed in conventional fully pumped heating systems. However, by using a combination of normally-open and normally-closed spring return motorized valves it is possible to utilize a solid fuel boiler in such a system. The Honeywell Y605A Panel Solid Fuel Timed Sundial Plan is a fully assembled, pre-plumbed and pre-wired control set and is designed specifically for use with a solid fuel boiler on a fully pumped system.

A feature of the Trianco TRG range of boilers is the use of an Economy Thermostat which is pre-set at a low temperature (57°C) for night operation and any day period when central heating is not required. A double circuit programmer is used to control the boiler. When hot water is programmed, the boiler operates on the Economy Thermostat. When central heating is selected, the boiler is controlled by the higher thermostat setting and the pump also starts (see Figure 7.1).

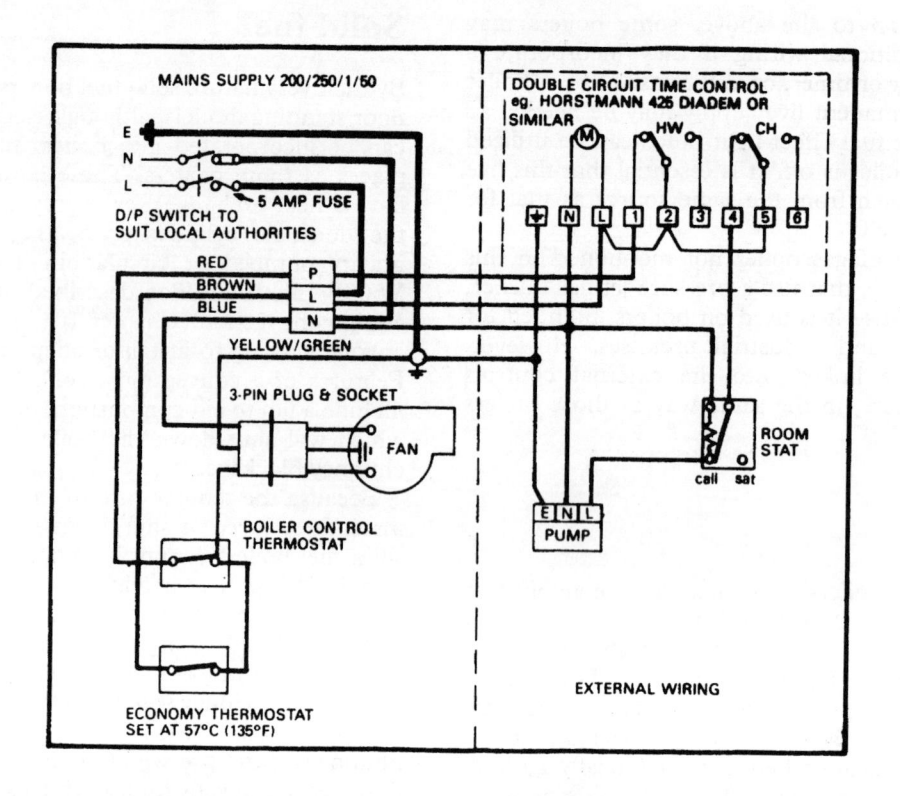

Installing and Servicing Domestic Heating Wiring Systems and Control

Figure 7.1

8

Boilers – gas

Ariston 20 MFS combination **Wall mounted**

Boiler is supplied with a 3-core mains lead.
External controls such as a time clock with voltage free contacts, or room thermostat suitable for use on 240V, should be connected as follows to the 12-way terminal strip:

```
TT    TR    TC    N    L
O     O     O ────O    O           Suitable for sealed systems
OUT   N     IN    Mains             Heat exchanger material, copper
```

Ariston DIA 20 MFFICE combination **Wall mounted**

Boiler is supplied with a 3-core mains lead.
A room thermostat suitable for use on low voltage and/or can be connected into terminals after removing either of the two brown links from the green connector on the PCB board (an external timer with voltage free terminals).

Suitable for sealed systems
Heat exchanger material, copper

AWB thermomaster **Wall mounted**
23.08 central heating only
23.09WT combination

```
O        O     O       O           Suitable for sealed systems
Do not         Rom Stat             Heat exchanger material, copper
 use            24V
```

Supply is via 3-core flex lead supplied fitted. Control circuit is 24V and room thermostat should be connected to block fitted on left-hand side of boiler.

AWB 45, 60, 75 **Floor or wall mounted**

```
E    N    L
O    O    O    O    O    O          Not suitable for use on sealed systems
E    N    Switch Int                 Heat exchanger material, steel
          Live
```

109

Installing and Servicing Domestic Heating Wiring Systems and Control

Barlo Balmoral

12	11	10		9	8		7		6		5	4	3	2	1
○	○	○		○	○		○		○		○	○	○	○	○
L	N	E		L	N		L								
Mains				Pump			Switch Live				Internal Wiring				

Wall mounted

Fully pumped systems only
Suitable for use on sealed systems
Heat exchanger material, copper

Terminals 7 and 12 are linked.

Barlow Blenheim
15/30, 30/40, 40/50, 50/60, 60/75

4	3		2		1		EP	LP	NP		N	L	E
○	○		○				○	○	○		○	○	○
Internal Wiring			Switch Live		I		E	L	N		N	L	E
							Pump				Mains		

Wall mounted

Fully pumped system only
Suitable for use on sealed systems
(kit required for 15/30)
Heat exchanger material, copper

Barlow Blenheim 42, 53

E	L	N		NP	LP	EP		1		2		3	4
○	○	○		○	○	○		○		○		○	○
E	L	N		Pump				L		Switch Live		Internal Wiring	
Mains													

Wall mounted

Fully pumped system only
Suitable for use on sealed systems
Heat exchanger material, copper

Barlo Duo Combination

L	N		1	2		3	4		5	6		E
○	○		○	○		○	○		○	○		○
L	N		N	L		Out	In		Out	In		E
Mains			Timer Motor			Timer Contacts			Room Stat Contacts			

Wall mounted

Suitable for sealed systems
Heat exchanger material, copper

If no timer or room thermostat link 3–4 and/or 5–6 as appropriate.

Baxi 401, 552

L	N	2
○	○	○
Switch Live	N	Perm* Live

Back boiler unit

Suitable for use on sealed systems
using optional kit
Heat exchanger material, cast iron

*Permanent live only required if fire front has light bulbs, e.g. B, GF Super, LFE3 Super, VP and SP fire fronts.

Baxi Boston

With/without integral programmer

N	L	1	2	3	4	L1	N1
○	○	○	○	○	○	○	○
N	L	HW Off	CH Off	HW On	CH On	Switch Live	N
Mains							

Floor standing

Not suitable for use on sealed systems
Heat exchanger material, cast iron

Programmer if fitted
Landis & Gyr RWB2

Boilers – gas

Baxi FS range

With/without integral programmer

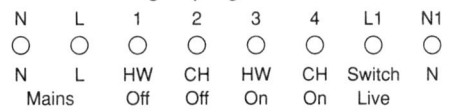

N	L	1	2	3	4	L1	N1
○	○	○	○	○	○	○	○
N Mains	L	HW Off	CH Off	HW On	CH On	Switch Live	N

Floor standing

Not suitable for use on sealed systems
Heat exchanger material, cast iron

Programmer if fitted
Landis & Gyr RWB2

Baxi Genesis combination

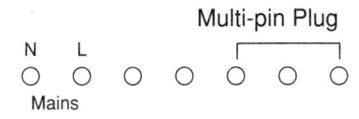

Multi-pin Plug

N	L					
○	○	○	○	○	○	○
Mains						

Wall mounted

Fully pumped systems only
Suitable for sealed systems
Heat exchanger material, copper

External controls, such as time clock with voltage free contacts, or room thermostat suitable for use at 24V, should be connected between terminals 5–7 after removing link.

Baxi Solo PF

PL	PN	PE	S/L	L	N	E
○	○	○	○	○	○	○
Pump			Switch Live	Mains		

Wall mounted

Fully pumped systems only
Suitable for use on sealed systems
Heat exchanger material, cast iron

Baxi Solo RS

L	N	E
○	○	○
Switch Live	N	E

Wall mounted

Not suitable for use on sealed systems
(see SOLO RS/SS)
Heat exchanger material, cast iron

Baxi Solo RS/SS

PL	PN	PE	S/L	L	N	E
○	○	○	○	○	○	○
Pump			Switch Live	Mains		

Wall mounted

Fully pumped systems only
Suitable for use on sealed systems
Heat exchanger material, cast iron

Baxi Solo 2 PF

	Pump			E	N	PL	SWL
○	○	○		○	○	○	○
L	N	E		E	N	Perm Live	Switch Live

Wall mounted

Fully pumped systems only
Suitable for use on sealed systems
Heat exchanger material, cast iron
Integral frost thermostat

Installing and Servicing Domestic Heating Wiring Systems and Control

Baxo Solo 2RS

Fully pumped systems

a) SL N E   PL L N E
   ○ ○ ○   ○ ○ ○ ○
   Switch N E   Perm Pump N E
   Live   Live Live

Wall mounted

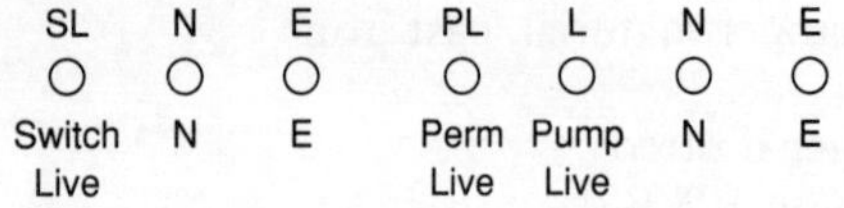

Suitable for use on sealed systems
Heat exchanger material, cast iron

Gravity hot water, pumped central heating systems

b) SL N E   PL L N E
   ○ ○ ○   ○ ○ ○ ○
   Switch N E   Not used
   Live

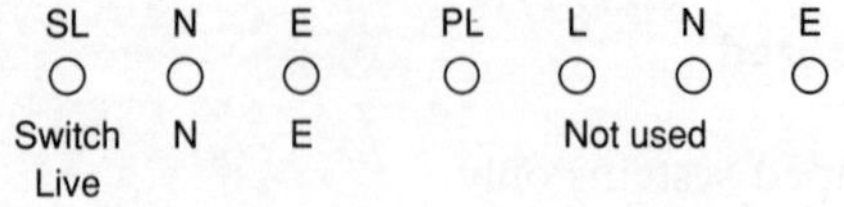

Suitable for use on sealed systems
Heat exchanger material, cast iron

When used on system b) the overheat thermostat should be by-passed by referring to installation instructions.

Baxi WM range

 L   N
 ○   ○   ○
Switch N
Live

Wall mounted

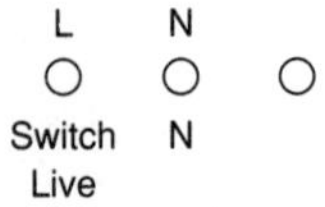

Not suitable for use on sealed systems
Heat exchanger material, cast iron

BKL Heatmaster combination

1 2   3 4 5
○ ○   ○ ○ ○
N L   Timer or N
        Room stat
        contacts

Wall mounted

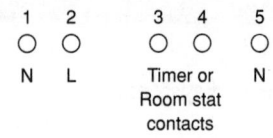

Suitable for sealed systems
Flue – conventional or fanned
If no timer or room thermostat link 3–4.

Boulter Centurion combination

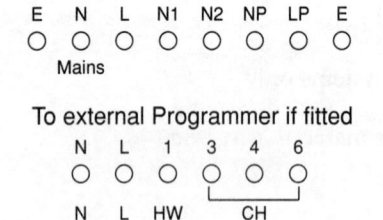

E N L N1 N2 NP LP E
○ ○ ○ ○   ○   ○   ○   ○
      Mains

To external Programmer if fitted
   N L 1   3 4 6
   ○ ○ ○   ○ ○ ○
   N L HW CH
          ON ON

Floor standing

Suitable for sealed systems
Heat exchanger material, copper

If no external controls link 4–6 only

Boulter Centurion WM 2.5 combination

L E N
○ ○ ○   ○ ○
   Mains

Wall mounted

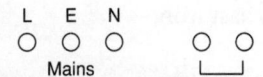

Suitable for sealed systems
Heat exchanger material, copper

Wire external controls, e.g. timer, room thermostat, into linked terminals. Remove link.

112

Boilers – gas

Boulter Gasray 2

L　N　3　4　5　6　7　8　9　10　11　12
○　○　○　○　○　○　○　○　○　○　○　○
Switch　N　N　　　　　　N　　　　N
Live

Floor standing

Fully pumped systems only
Suitable for use on sealed systems
Heat exchanger material, cast iron

Centurion Combi

See Boulter Centurion

Chaffoteaux Britony combination

Multi-pin Plug
N　　Ph
○　○　○　○　○　○　○
N　L

Wall mounted

Suitable for sealed systems
Heat exchanger material, copper

Connect room thermostat suitable for 24V into multiplying plug removing link.

Chaffoteaux Celtic combination

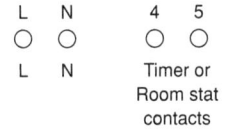

L　N　　4　5
○　○　　○　○
L　N　　Timer or
　　　　Room stat
　　　　contacts

Wall mounted

Suitable for sealed systems
Heat exchanger material, copper
Integral frost thermostat fitted

If no timer or room thermostat fit link 4–5.

Chaffoteaux Celtic Plus FF combination

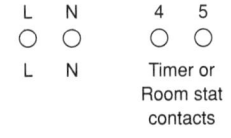

L　N　　4　5
○　○　　○　○
L　N　　Timer or
　　　　Room stat
　　　　contacts

Wall mounted

Suitable for sealed systems
Heat exchanger material, copper

If no timer or room thermostat fit link 4–5.

Chaffoteaux Challenger

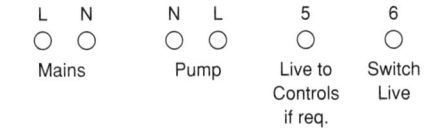

L　N　　N　L　　5　　　6
○　○　　○　○　　○　　○
Mains　　Pump　Live to　Switch
　　　　　　　Controls　Live
　　　　　　　if req.

Wall mounted

Fully pumped systems only
Suitable for use on sealed systems
Heat exchanger material, copper

Chaffoteaux Flexiflame 140   Wall mounted

```
L    N      7    6    8
O    O      O    O    O
Switch  N   Timer or   N
Live        Room stat
            contacts
```

Fully pumped systems only
Suitable for use on sealed systems
Heat exchanger material, copper
The integral pump is not the system pump

If no timer or room thermostat fit link 6–7.

Chaffoteaux Sterling OF and FF combination   Wall mounted

```
L1   N2     3    4    5
O    O      O    O    O
  Mains
```

Suitable for use on sealed systems
Heat exchanger material, copper

External controls such as time clock with voltage free contacts, or room thermostat, should be connected between terminals 4–5. Terminal 3 can be used to connect room thermostat neutral.

Chaffoteaux Sterling SB system boiler   Wall mounted

Available with or without 3-port mid-position valve.

Without valve – heating only:

```
   Room Stat         N    L    E
O    O    O         O    O    O
N   COM  DEM          Mains
```

Suitable for use on sealed systems
Heat exchanger material, copper

Link N–N, COM–L only

With 3-port mid-position valve:
Connect valve as per colours indicated on terminal strip. Room thermostat and mains as above.

```
        Programmer            Cylinder Stat
N   L   5    4   3   2   1    7    0    1
O   O   O    O   O   O   O    O    O    O
  Mains     CH       ON       SAT COM DEM
            ON       ON
```

Suitable for use on sealed systems
Heat exchanger material, copper

Link L–L, N–N, N–blue. Connections: White–room stat demand, orange–Cyl stat satisfied Prog 4–room stat common, prog 1–cyl stat common.

Combi Company GEM combination   Wall mounted

```
L    N    E       L2   L1
O    O    O       O    O
  Mains          12V DC
```

Suitable for use on sealed systems
Heat exchanger material, copper

External controls such as a time clock with voltage free contacts or room thermostat should be connected between terminals L2–L1. A frost thermostat should be wired across L2 and red wire connected to DHW flow switch.

Boilers – gas

ELM Leblanc GLM 5.20, 5.32 combination

Wall mounted

N   L
○ ○   ○ ○   ○ ○
240V   Timer   Room stat
Mains

Suitable for sealed systems
Heat exchanger material, copper
The timer should have voltage free terminals. The room thermostat should be suitable for low voltage (32V).

ELM Leblanc GVM 4.20 combination

Wall mounted

N   L
○ ○   ○ ○   ○ ○
240V   Timer   Room stat
Mains

Suitable for sealed systems
Heat exchanger material, copper
The timer should have voltage free contacts. Room thermostat should be suitable for mains voltage 240V.

ELM Leblanc GVM 7.23 combination   GVM 7.28

Wall mounted

○ ○   ○ ○
Timer   Room stat

○   ○   ○
Mains 240V

Suitable for sealed systems
Heat exchanger material, copper
Timer and or room thermostat should be connected into their appropriate terminals and be suitable for 24V.

ELM Leblanc GVM 14/20 combination

Wall mounted

N   L
○ ○   ○ ○   ○ ○
240V   Timer   Room stat
Mains

Suitable for sealed systems
Heat exchanger material, copper

ELM Leblanc GVM C.21, C.23 condensing combination

Wall mounted

N   L
○ ○   ○ ○   ○ ○
240V   Timer   Room stat
Mains

Suitable for sealed systems
Heat exchanger material, copper

Timer should have voltage free contacts. Room thermostat should be suitable for low voltage 32V.

Eurocombi Styx combination

Wall mounted

Boiler is supplied with a 3-core mains lead
External controls such as time clock with voltage free contacts or room thermostat should be connected into connector C after removing either brown link.

Suitable or sealed systems.
Heat exchanger material, copper

Installing and Servicing Domestic Heating Wiring Systems and Control

Ferroli 76FF combination

```
 24V          L  N  E
 O  O         O  O  O
Timer or       Mains
Room stat
```

Wall mounted

Suitable for sealed systems
Heat exchanger material, copper
Room thermostat should be suitable for 24V.

If no timer or room thermostat fits link. Do not wire room thermostat neutral.

Ferroli 77 CF, FF and FF Popular combination

```
 24V          E  N  L
 O  O         O  O  O
Timer or       Mains
Room stat
```

Wall mounted

Suitable for sealed systems
Heat exchanger material, copper
77FF Popular does not include integral clock.

Ferroli 100 FF, 120 CF combination

```
 L  E  N      4  5
 O  O  O      O  O
  Mains       24V
```

Wall mounted

Suitable for sealed systems
Heat exchanger material, copper

External controls such as time clock with voltage free contacts, or room thermostat suitable for use at 24V, should be connected between terminals 4–5 after removing link.

Ferroli Logica condensing combination

```
 L  E  N      4  5
 O  O  O      O  O
  Mains       24V
```

Wall mounted

Suitable for sealed systems
Heat exchanger material, copper

External controls such as time clock with voltage free contacts, or room thermostat suitable for use at 24V, should be connected between terminals 4–5 after removing link.

Ferroli Nouvelle Elite 127/92 combination

```
 L  E  N      4  5
 O  O  O      O  O
Mains 240V    24V
              Room stat
```

Wall mounted

Suitable for sealed systems
Heat exchanger material, copper

External controls such as time clock, or room thermostat suitable for use at 24V, should be connected between terminals 4–5 after removing link. Do not wire room thermostat neutral.

Boilers – gas

Ferroli Optima combination

```
L   E   N      4   5
O   O   O      O   O
   Mains         24V
```

Wall mounted

Suitable for sealed systems
Heat exchanger material, copper

External controls such as time clock with voltage free contacts, or room thermostat suitable for use at 24V, should be connected between terminals 4–5 after removing link.

Ferroli Roma

```
 1      2SL     L   N   E     E  3PN  4PL
 O       O      O   O   O     O   O    O
Live to Switch    Mains           Pump
Controls Live in
if req.
```

Wall mounted

Fully pumped systems only
Suitable for use on sealed systems
Heat exchanger material, copper
Integral frost thermostat fitted

Ferroli Xignal combination

```
1   2   3   4   5   6   7   8   9   10
O   O   O   O   O   O   O   O   O   O
L   E   N       Room    External
  Mains         Sensor   Sensor
```

Wall mounted

Suitable for sealed systems
Heat exchanger material, copper
Both sensors are supplied with the boiler. Computerized boiler with synthesized voice communication.

Firefly Sole Provider SP80ER and SP80EF combination

Mains lead supplied.
If fitting room thermostat remove link and connect across link terminals.

Wall mounted

Suitable for sealed systems
Heat exchanger material, copper

Gemini 960E combination

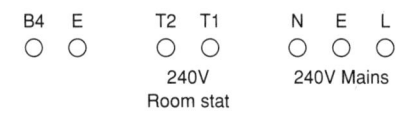

```
B4   E      T2   T1      N   E   L
O    O      O    O       O   O   O
            240V         240V Mains
            Room stat
```

Wall mounted

Suitable for sealed systems
Heat exchanger material, copper

Electrical connection via 7-pin plug as shown. Connect frost thermostat across B4–T1 in plug.

Geminox MZ condensing

```
P   P       TH   TH      N   N   PH
O   O       O    O       O   O   O
Link        Timer or     N   Mains
            Room stat
            contacts
```

Wall mounted

Fully pumped systems only
Suitable for sealed systems
Heat exchanger material, aluminium

Link P–P to be removed if fitting digi control.

Glotec GT80 condensing

Wall mounted

10	9	8	7	6	5	4	3	2	1
○	○	○	○	○	○	○	○	○	○
Switch Live	Live to contacts if req.	N	E	E	L	N	N	L	E
		Pump (optional connection)					Mains		

Fully pumped systems only
Suitable for use on sealed systems
Heat exchanger material, stainless steel

Glow-Worm 45 and 56B

Back boiler unit

N	L	N	SL
○	○	○	○
N	L*	N	Switch Live

Not suitable for use on sealed systems
Heat exchanger material, cast iron

*Permanent live only required if fire-front has light bulbs, e.g. Homeglow and LFC

Glow-Worm Economy

Wall mounted

7	8	9	10	11	12
○	○	○	○	○	○
L	N	L	N	L	Switch Live
	Pump		Mains		

Remove link 7–12

Fully pumped systems only
Suitable for use on sealed systems
Heat exchanger material, copper

Glow-Worm Energysaver condensing

Wall mounted

N	L	1	PN	PL	SL
○	○	○	○	○	○
Mains			Pump		Switch Live

Remove link 7–12

Fully pumped systems only
Suitable for use on sealed systems
Heat exchanger material, copper

Glow-Worm Express combination

Wall mounted

5-pin internal plug

L	N	E		
○	○	○	○	○
L	N	E	Timer or Room stat contacts	
	Mains			

Suitable for sealed systems
Heat exchanger material, copper

Glow-Worm Fuelsaver 'B'

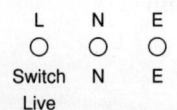

Wall mounted

L	N	E
○	○	○
Switch Live	N	E

Fully pumped systems only
Suitable for use on sealed systems
Heat exchanger material, copper

Boilers – gas

Glow-Worm Fuelsaver 'B' MK2

```
 7    8    9   10   11   12
 O    O    O    O    O    O
 L    N    L    N    L   Switch
Pump      Mains         Live
```

Remove link 7–12

Wall mounted

Fully pumped systems only
Suitable for use on sealed systems
Heat exchanger material, copper

Glow-Worm Fuelsaver 'BR' MK2

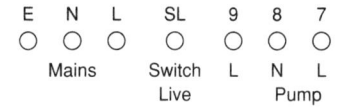

```
 E    N    L    SL   9    8    7
 O    O    O    O    O    O    O
    Mains       Switch  L    N    L
                Live        Pump
```

Remove link SL–9

Wall mounted

Fully pumped systems only
Suitable for use on sealed systems
Heat exchanger material, copper

Glow-Worm Fuelsaver Complhead condensing

Electrical connection of switch live (L), Neutral (N), and earth (E), via 3-pin internal plug

Wall mounted

Fully pumped systems only
Suitable for sealed systems
Heat exchanger material, copper

Glow-Worm Fuelsaver 'F'

```
 7    8    9   10   11   12
 O    O    O    O    O    O
 L    N    L    N    L   Switch
Pump      Mains         Live
```

Remove link 7–12

Wall mounted

Fully pumped systems only
Suitable for use on sealed systems except 100F
Heat exchanger material, copper

Glow-Worm Fuelsaver MK2

```
 7    8    9   10   11   12
 O    O    O    O    O    O
 L    N    L    N    L   Switch
Pump      Mains         Live
```

Remove link 7–12

Wall mounted

Fully pumped systems only
Suitable for use on sealed systems except 75 MK2
Heat exchanger material, copper

Glow-Worm Fuelsaver 'R' MK2

```
 E    N    L    SL   9    8    7
 O    O    O    O    O    O    O
 E    N    L   Switch  L    N    L
    Mains      Live        Pump
```

Remove link SL–9

Wall mounted

Fully pumped systems only
Suitable for use on sealed systems except 75R MK2
Heat exchanger material, copper

119

Installing and Servicing Domestic Heating Wiring Systems and Control

Glow-Worm Fuelsaver UFB

```
E   N   L   SL   9   8   7
O   O   O   O    O   O   O
E   N   L   Switch  L    N    L
Mains       Live        Pump
```

Remove link SL–9

Wall mounted

Fully pumped systems only
Suitable for use on sealed systems except 100F
Heat exchanger material, copper

Glow-Worm Hideaway

Electrical connection via 3-pin internal plug

Floor standing

Not suitable for use on sealed systems
Heat exchanger material, cast iron

Glow-Worm Spacesaver 75 CF

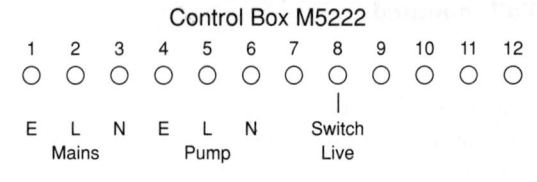

```
            Control Box M5222
1   2   3   4   5   6   7   8   9   10  11  12
O   O   O   O   O   O   O   O   O   O   O   O
                            |
E   L   N   E   L   N       Switch
Mains       Pump            Live
```

Remove link 7–8

Wall mounted

Fully pumped systems only
Not suitable for use on sealed systems
Heat exchanger material, cast iron

Glow-Worm Spacesaver 'B' MK2

```
7     8        9   10  11    12
O     O        O   O   O     O
Spare Internal     E   N     Switch
      Wiring                 Live
```

Wall mounted

Not suitable for use on sealed systems
Heat exchanger material, cast iron

Glow-Worm Spacesaver 'BR' MK2

```
E   N   L
O   O   O
E   N   Switch
        Live
```

Wall mounted

80BR MK2 – Fully pumped systems only
Not suitable for use on sealed systems
Heat exchanger material, cast iron

Glow-Worm Spacesaver 'F'

```
L        N   E
O        O   O
Switch   N   E
Live
```

Wall mounted

Not suitable for use on sealed systems
Heat exchanger material, cast iron

Glow-Worm Spacesaver 'KFB'

```
5    6    7    8    9   10   11   12
O    O    O    O    O    O    O    O
E    E    L    N    L    N    L   Switch
              Pump      Mains     Live
```

Remove link 7–12

Wall mounted

Fully pumped systems only
Not suitable for use on sealed systems
Heat exchanger material, cast iron

Glow-Worm Spacesaver 'R' MK2

```
E    N    L
O    O    O
E    N   Switch
         Live
```

Wall mounted

Not suitable for use on sealed systems
Heat exchanger material, cast iron

Glow-Worm Spacesaver 'RF'

```
L    N    E
O    O    O
Switch N    E
Live
```

Wall mounted

Not suitable for use on sealed systems
Heat exchanger material, cast iron

Glow-Worm Swiftflow combination

Electrical connections are via a 5-pin plug

```
L    N    E         2    1
O    O    O         O    O
L    N    E         L    L
   Mains           Out   In
```

Wall mounted

Suitable for sealed systems
Heat exchanger material, copper

External controls can be connected between terminal 2–1 or directly into terminal 1 after removing link.

Glow-Worm Ultimate BF and CF

```
L    N    E
O    O    O
Switch N    E
Live
```

Wall mounted

80 BF – Fully pumped systems only
Not suitable for use on sealed systems
Heat exchanger material, cast iron

Glow-Worm Ultimate BFSS

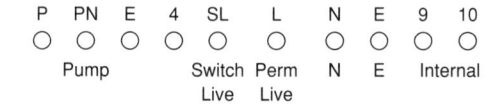

```
P   PN   E    4   SL    L    N    E    9   10
O    O   O    O    O    O    O    O    O    O
  Pump        Switch Perm  N    E  Internal
              Live   Live
```

Wall mounted

Fully pumped systems only
Suitable for use on sealed systems
Heat exchanger material, cast iron

Remove link 4–SL if wiring external controls

Glow-Worm Ultimate FF
(excl. 80FF – see below)

Gravity hot water, pumped central heating

E	N	L	SL	9	PN	PL	E	K1	K2
○	○	○	○	○	○	○	○	○	○
E	N		Switch				Link		
			Live						

Fully pumped systems

E	N	L	SL	9	PN	PL	E	K1	K2
○	○	○	○	○	○	○	○	○	○
Mains		Switch		Pump			No		
		Live					Link		

Wall mounted

Suitable for use on sealed systems
Heat exchanger material, cast iron

Glow-Worm Ultimate 80FF

E	7	8	9	SL	L	N	E
○	○	○	○	○	○	○	○
E	L	N		Switch	Perm	N	E
Pump				Live	Live		

Remove link 9–SL when wiring external controls

Wall mounted

Fully pumped systems only
Not suitable for use on sealed systems
Heat exchanger material, cast iron

Halstead 40H–50H

E	L	N	NP	LP	EP	1	2	3	4
○	○	○	○	○	○	○	○	○	○
E	L	N		Pump		L	Switch	Internal	
Mains							Live	Wiring	

Wall mounted

Fully pumped systems only
Not suitable for use on sealed systems
Heat exchanger material, copper

Halstead 45F, 65F

12	11	10	9	8	7	6	5	4	3	2	1
○	○	○	○	○	○	○	○	○	○	○	○
L	N	E	L	N		L	Switch	Internal Wiring			
Mains			Pump				Live				

Terminals 7 and 12 are linked.

Wall mounted

Fully pumped systems only
Suitable for use on sealed systems
Heat exchanger material, copper

Halstead Balmoral

12	11	10	9	8	7	6	5	4	3	2	1
○	○	○	○	○	○	○	○	○	○	○	○
L	N	E	L	N		L	Switch	Internal Wiring			
Mains			Pump				Live				

Terminals 7 and 12 are linked.

Wall mounted

Fully pumped systems only
Suitable for use on sealed systems
Heat exchanger material, copper

Boilers – gas

Halstead Bentley

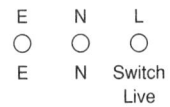

E	N	L
○	○	○
E	N	Switch Live

Wall mounted

Suitable for use on sealed systems using optional kit
Heat exchanger material, cast iron

Halstead Best

See Halstead Boss – renamed

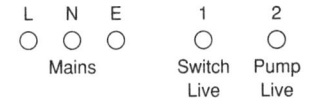

Halstead Blenheim

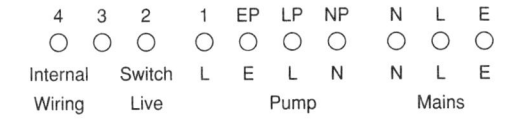

4	3	2	1	EP	LP	NP	N	L	E
○	○	○	○	○	○	○	○	○	○
Internal Wiring		Switch Live		L Pump	E	L	N Mains	L	E

Wall mounted

Fully pumped systems only
Suitable for use on sealed systems (15/30 using optional kit)
Heat exchanger material, copper

Halstead Boss

L	N	E	1	2
○	○	○	○	○
Mains			Switch Live	Pump Live

Wall mounted

Suitable for use on sealed systems
Heat exchanger material, cast iron

A permanent live and pump live must be wired into the boiler for all systems including gravity hot water. The plug at the front of the control box should be positioned as necessary.

Halstead Buckingham and Buckingham 2

E	N	L1
○	○	○
E	N	Switch Live

Floor standing

Not suitable for use on sealed systems
Heat exchanger material, cast iron

Halstead Quattro combination

L	N	E	1	2
○	○	○	○	○
Mains			Room Stat	

Wall mounted

Suitable for sealed systems
Heat exchanger material, copper
Integral-frost thermostat fitted

Room thermostat should be suitable for 240V.

Halstead Quattro Gold combination

```
L   N   E       1   2
O   O   O       O   O
   Mains         Room
                  Stat
```

Wall mounted

Suitable for sealed systems
Heat exchanger material, copper
Integral-frost thermostat fitted

Room thermostat should be suitable for 240V.

Halstead Trio combination

Connection via internal plug
```
L   N   E       41  40
O   O   O       O   O
   Mains       Room stat 24V
```

Wall mounted

Suitable for sealed systems
Heat exchanger material, copper

If no room thermostat link 40–41. Time clock integral to boiler.

Ideal Cavalcade

Electrical connection via 3-pin plug

Back boiler unit

Not suitable for use on sealed systems
Heat exchanger material, cast iron

Ideal Classic BF

```
N   E   L
O   O   O   O
N   E  Switch Int.
```

Wall mounted

Suitable for use on sealed systems
using optional kit
Heat exchanger material, cast iron

Ideal Classic FF

Electrical connection via 3-pin plug

Wall mounted

Not suitable for use on sealed systems
using optional kit
Heat exchanger material, cast iron

Ideal Classic LX

As Classic with glass front panel and enhanced casing trim.

Ideal Classic system boiler

Electrical connection via 3-pin plug

Wall mounted

Fully pumped systems only
Suitable for sealed systems
Heat exchanger material, cast iron

Boilers – gas

Ideal Classic Combi NF80 combination

CH RS1 RS2 L1 CL HW   E   L   N   CN
○   ○   ○   ○   ○   ○   ○   ○   ○   ○
    Room   L   L   Mains   N
    Stat

Wall mounted

Suitable for sealed systems
Heat exchanger material, cast iron

Fit link L1–HW. Connect room thermostat between terminals RS1 out – RS2 (in) after removing link. Connect external timer between terminals CL (live – CH (control). Terminal CN is for room thermostat and timer neutral. A programmer can be installed as above with hot water channel connected to HW after removing link L1–HW.

Ideal Compact Extra system boiler

L N E   PN PL   DVN DVL   E   1   2   A   B
○ ○ ○   ○   ○   ○   ○   ○   ○   ○   ○   ○
Mains   Pump   Diverter Valve   Room   Cylinder
supplied   Integral Integral   Sensor   Sensor

Wall mounted

Suitable for use on sealed systems
using optional kit
Heat exchanger material, copper

Boiler incorporates a priority valve and a link to alter from HW to CH priority if required. Sensors could be wired in bell wire.

Ideal Elan CF, RS

LG NG   LP NP   L   N
○   ○   ○   ○   ○   ○
Gas Valve   Optional pump Switch N
              connection   Live

Wall mounted

Fully pumped systems only
Suitable for use on sealed systems
Heat exchanger material, copper
See also Ideal Elan 2NF, RS

Ideal Elan 2CF, F, RS

LG NG   LP NP   L   N
○   ○   ○   ○   ○   ○
Gas Valve   Optional pump Switch N
              connection   Live

Wall mounted

Fully pumped systems only
Suitable for use on sealed systems
Heat exchanger material, copper

Ideal Elan 2CF, NF, RS

○   ○   ○   ○   ○
Link for   Switch E   N
Optional   Live
Overheat
Thermostat

Wall mounted

Fully pumped systems only
Suitable for use on sealed systems
Heat exchanger material, copper
See also Elan 2CF, F, RS

Installing and Servicing Domestic Heating Wiring Systems and Control

Ideal Excel

L	N	LB	NP	LP	M	S
○	○	○	○	○	○	○
Mains Live		Switch	Pump		*HW Valve Motor	*Cyl Stat Demand

Wall mounted

Fully pumped systems only
Not suitable for use on sealed systems
Heat exchanger material, cast iron

*If used in conjunction with one 2-port in heating circuit only or one 3-port mid-position valve wire as shown excluding connections to M and S. However if wiring with two 2-port motorized valves, terminals M and S are wired as shown and the auxiliary switch wires of the hot water motorized valve are not connected.

Ideal Mexico Slimline

Electrical connection via 3-pin internal plug

Floor standing

Not suitable for use on sealed systems
Heat exchanger material, cast iron

Ideal Mexico Slimline 2

Electrical connection via 3-pin internal plug

Floor standing

Not suitable for use on sealed systems
Heat exchanger material, cast iron

Ideal Mexico Super

N	E	L2	L1	L	E	N	L3	E	N
○	○	○	○	○	○	○	○	○	○
N	E	*	*	Switch Live	E	N		E	N

Floor standing

Not suitable for use on sealed systems
Heat exchanger material, cast iron

*If convenient to do so, the pump connection can be made in L2 for gravity systems or L1 for fully pumped systems

Ideal Mexico Super 2

N	E	LP	LP	LB	E	N	LG	E	N
○	○	○	○	○	○	○	○	○	○
N	E	*	*	Switch Live	E	N	Internal wiring to gas valve		

Floor standing

Not suitable for use on sealed systems
Heat exchanger material, cast iron

*Terminals LP are provided for connecting pump if convenient to do so.

Ideal Minimiser FF condensing

A switched live and a neutral are connected into the 2-terminal connector block.

Wall mounted

Fully pumped systems only
Suitable for use on sealed systems
Heat exchanger material, aluminium

Boilers – gas

Ideal Response FF combination

```
E   N   L       R1  R2
O   O   O       O   O
    Mains       └───┘
```

Wall mounted

Suitable for sealed systems
Heat exchanger material, copper and cast iron

External controls such as time clock with voltage free contacts, or room thermostat, should be connected between terminals R1 – R2 after removing link. Room thermostat neutral can be connected to N.

Ideal Sprint 80F combination

```
LP  N   E   L1  N   L2  L0      L   E   N
O   O   O   O   O   O   O       O   O   O
  Pump          N                 Mains
```

Wall mounted

Suitable for sealed systems
Heat exchanger material, copper

Ideal Sprint RS75 combination

```
LP  N   E   L1  N   L2  L0      L   E   N
O   O   O   O   O   O   O       O   O   O
  Pump          N                 Mains
```

Wall mounted

Suitable for sealed systems
Heat exchanger material, copper

External controls such as time clock with voltage free contacts, or room thermostat, should be connected between terminals L1–L2. Connect frost thermostat across L0–L2. Thermostats should be suitable for 240V.

Ideal Spring Rapide RS 75N combination

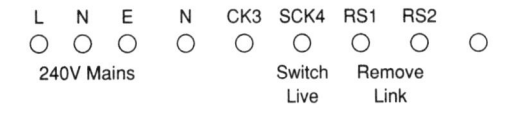

```
L   N   E       N   CK3 SCK4    RS1 RS2
O   O   O       O   O   O       O   O   O
 240V Mains         Switch  Remove
                     Live    Link
```

Wall mounted

Suitable for sealed systems
Heat exchanger material, Copper

Connect frost thermostat across FRS1–FRS2.

Ideal Sprint Rapide 90NF combination

(a) Permanent mains connection required
(b) Connect room thermostat between RS1–RS2
(c) Connect timer switch live to SCK4 (remove link CK3–SCK4)
(d) Connect frost thermostat between FRS1–FRS2

Wall mounted

Suitable for sealed systems
Heat exchanger material, copper

Ideal Turbo 2 condensing

```
                              CH   HW   CH   HW
LP   NP   LB    N    L   NT   ON   ON   OFF  OFF
O    O    O     O    O   O    O    O    O    O
Optional Switch N    L   N    Programmer Terminals
Pump     Live  Mains*          if fitted
```

Wall mounted

Fully pumped systems only
Suitable for use on sealed systems
Heat exchanger material, aluminium

*Permanent live is only required if internal programmer fitted

Ideal W2000

Electrical connection via 3-pin internal plug

Wall mounted

Suitable for use on sealed systems using optional kit
Heat exchanger material, cast iron

Ideal WCF, WRS

Electrical connection via 3-pin internal plug

Wall mounted

Not suitable for use on sealed systems
Heat exchanger material, cast iron

Ideal WLX

Electrical connection via 3-pin internal plug

Wall mounted

Not suitable for use on sealed systems
Heat exchanger material, cast iron

Keston 50, 60, 80 condensing

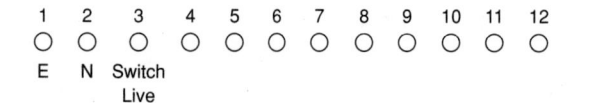

```
1    2    3    4    5    6    7    8    9    10   11   12
O    O    O    O    O    O    O    O    O    O    O    O
E    N    Switch
          Live
```

Wall mounted

Fully pumped systems only
Suitable for use on sealed systems
Heat exchanger material, stainless steel

Malvern 30–70 condensing

```
6    5    4    3    1    N    N    N    L    L    E    E
O    O    O    O    O    O    O    O    O    O    O    O
              Switch          N   Perm            E
              Live                Live
```

Wall mounted

Fully pumped systems only
Suitable for use on sealed systems
Heat exchanger material, copper, aluminium

Maxol EM25

```
     L    E    N
     O    O    O
 Switch   E    N
 Live
```

Wall mounted

Fully pumped systems only
Not suitable for use on sealed systems
Heat exchanger material, copper

Boilers – gas

Maxol EM40, EM50

E	L	N	NP	LP	EP	1	2	3	4
○	○	○	○	○	○	○	○	○	○
E	L	N		Pump			L	Switch	Internal
Mains								Live	Wiring

Wall mounted

Fully pumped systems only
Not suitable for use on sealed systems
Heat exchanger material, copper

Maxol Homewarm 600

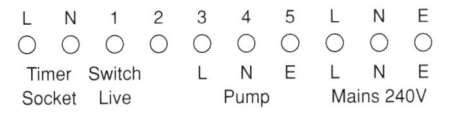

L	N	1	2	3	4	5	L	N	E
○	○	○	○	○	○	○	○	○	○
Timer	Switch		L	N	E	L	N	E
Socket	Live			Pump			Mains 240V

Terminals L–2–L are linked. Terminals N–4–N are linked.

Wall mounted

Fully pumped systems only
Not suitable for use on sealed systems
Heat exchanger material, copper

Maxol Microturbo 40

E	1	2	3	4	5	L	N	E
○	○	○	○	○	○	○	○	○
E	Switch		L	N	E	L	N	E
	Live			Pump			Mains 240V

Terminals 2–L Mains are linked.

Wall mounted

Fully pumped systems only
Suitable for use on sealed systems
Heat exchanger material, copper

Maxol Morocco

L	N	1	2	3	4	5	L	N	E
○	○	○	○	○	○	○	○	○	○
Timer	Switch		L	N	E	L	N	E
Socket	Live			Pump			Mains 240V

Terminals L–2–L are linked. Terminals N–4–N are linked.

Wall mounted

Fully pumped systems only
Suitable for use on sealed systems
Heat exchanger material, copper

Maxol Mystique

A	B	1	2	3	4	5	L	N	E
○	○	○	○	○	○	○	○	○	○
Spare	N	Switch		L	N	E	L	N	E
		Live			Pump			Mains 240V

Terminals 2–L are linked.

Wall mounted

Fully pumped systems only
Not suitable for use on sealed systems
Heat exchanger material, copper

Myson Apollo

		HW	CH	CH			
N	L	ON	ON	OFF	ON	N	L
○	○	○	○	○	○	○	○
N	L	Programmer		Switch	N	L
Mains		Terminals		Live	Pump
		if fitted

Wall mounted

Fully pumped systems only
Suitable for use on sealed systems
Heat exchanger material, copper

Myson Gemini combination

```
E   N   L   R0 R1
O   O   O   OO
Mains 240V  Room stat
```

Wall mounted

Suitable for use on sealed systems using optional kit
Heat exchanger material, copper

Myson (Thorn) Housewarmer

Back boiler unit

Gravity hot water, pumped central heating or fully pumped without overheat kit fitted

```
L      N   1   2   3   4
O      O   O   O   O   O
Switch N
Live
```

Not suitable for use on sealed systems
Heat exchanger material, cast iron

Terminals 1–4 could be used for room thermostat, pump, etc., *or*

Fully pumped with overheat kit fitted for pump over-run

```
L       N   1      2      3     4
O       O   O      O      O     O
Switch  N   Perm  Spare  Pump   N
Live        Live         Live
```

Myson Housewarmer 2

Back boiler unit

```
L1      L       N   E
O       O       O   O
*Perm   Switch  N   E
Live    Live
```

Not suitable for use on sealed systems
Heat exchanger material, cast iron
*Permanent live only required for fire front bulbs if fitted.

Myson Housewarmer electronic

Back boiler unit

```
L1      L       N   E   Int
O       O       O   O   O
*Perm   Switch  N   E
Live    Live
```

Not suitable for use on sealed systems
Heat exchanger material, cast iron
*Permanent live only required for fire front bulbs if fitted.

Myson Marathon (excluding 1500C)

Floor standing

```
E   N   L   1    2    3    L       N   E   4
O   O   O   O    O    O    O       O   O   O
E   N   L   HW   CH   Switch L     N   E   Internal
Mains       ON   ON   Live   *Pump         Wiring
```

Not suitable for use on sealed systems
Heat exchanger material, cast iron
*Alternative pump connection

Permanent live only required if internal programmer fitted.

Myson Marathon 1500C

E	N	L	1	2	3	L	N	E	4
○	○	○	○	○	○	○	○	○	○
E	N	L	Switch			L	N	E	
Mains			Live			Pump			

Remove link L–1 and 2–Pump L.

Floor standing

Not suitable for use on sealed systems
Heat exchanger material, cast iron
Fit blue plug for gravity hot water
Fit red plug for fully pumped

Myson Midas B combination

E2	N2	R1	R0	N	L	E
○	○	○	○	○	○	○
Room stat				Mains		

Remove Link R1–R0 if wiring room thermostat.

Wall mounted

Suitable for sealed systems
Heat exchanger material, copper

Myson Midas Si combination

L	E	N	R1	R0	Int.
○	○	○	○	○	○
Mains			Room stat		
240V			240V		

Wall mounted

Suitable for use on sealed systems
Heat exchanger material, copper

Connect room thermostat across terminals R1-R0 after removing link. **Do not** connect room thermostat neutral. If wiring a frost thermostat, connect across terminals N and R1.

Myson Olympic

N	L	1	2	3	4	5	L	N
○	○	○	○	○	○	○	○	○
N	L	Switch	CH	CH	HW	Spare	L	N
*Pump		Live	ON	OFF	ON		Mains	
			Programmer					
			Terminals if fitted					

Permanent live only required if internal programmer fitted or fully pumped systems
Fit blue plug for gravity hot water
Fit red plug for fully pumped
*Optional pump connection, remove all links if programmer fitted.

Wall mounted

Not suitable for use on sealed systems
Heat exchanger material, cast iron

A fan timer kit may be fitted to the Olympic, causing the fan to run for approximately 1 minute in every 16 minutes. The wiring of the boiler may be re-arranged to allow for this.

Myson Orion

N	L	1	2	3	4	5	L	N
○	○	○	○	○	○	○	○	○
N	L	Switch	CH	Spare	HW		L	N
Mains		Live	ON		ON		Mains if	
							Programmer fitted	

Remove links L–2, 1–4, 4–5 and 2–4 if programmer fitted.

Wall mounted

75 Si only suitable for use on sealed systems. Others: suitable for use on sealed systems using optional kit.
Heat exchanger material, cast iron

Ocean Alpha 240/280

```
L    N    E    1    2
O    O    O    O    O
Mains      L    L
           In   Out
```

Wall mounted

Fully pumped systems only
Suitable for sealed systems
Heat exchanger material, copper/stainless steel

External controls such as time clock with voltage free contacts, or room thermostat suitable for use at 240V, should be connected between terminals 1–2 after removing link.

Ocean 80, OF, FF, FF style combination

```
E    L    N    2    1
O    O    O    O    O
   Mains 240V
```

Wall mounted

Suitable for sealed systems
Heat exchanger material, copper

External controls such as time clock with voltage free contacts, or room thermostat suitable for use at 240V, should be connected between terminals 1–2 after removing link. Room thermostat neutral can be connected to terminal N.

Potterton Envoy

```
L    N    SWL     E    PL     1     2     3
O    O    O       O    O      O     O     O
L    N    Switch  E    Pump   CH    HW    HW
Mains     Live         Live   ON    ON    OFF
```

Wall mounted

Fully pumped systems only
Suitable for use on sealed systems
Heat exchanger material, aluminium
Integral frost thermostat

Terminals 1, 2, 3, only used if optional timer used.

Potterton Fireside

```
L        N    E
O        O    O
Switch   N    E
Live
```

Back boiler unit

Not suitable for use on sealed systems
Heat exchanger material, cast iron

Potterton Flamingo

```
E    N    L
O    O    O
E    N    Switch
          Live
```

Wall mounted

Not suitable for use on sealed systems
Heat exchanger material, cast iron

Potterton Flamingo 2

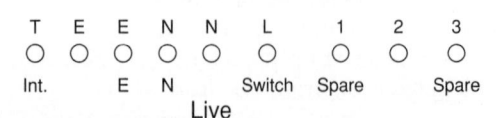

```
T    E    E    N    L    1     2     3
O    O    O    O    O    O     O     O
Int.      E    N    Switch Spare    Spare
               Live
```

Wall mounted

Not suitable for use on sealed systems
Heat exchanger material, cast iron

Terminals L–2 are linked.

Boilers – gas

Potterton Flamingo 3

Gravity hot water, pumped central heating

L	N	E	SWL	L	N	E
○	○	○	○	○	○	○
			Switch Live		N	E

Fully pumped systems

L	N	E	SWL	L	N	E
○	○	○	○	○	○	○
Pump			Switch Live	Perm Live	N	E

Wall mounted

Suitable for use on sealed systems
Heat exchanger material, cast iron

Potterton Housewarmer 45, 55

Gravity hot water, pumped central heating

N	L*	SWL	E	N	L
○	○	○	○	○	○
Mains		Switch Live	E		

Fully pumped systems

N	L	SWL	E	N	L
○	○	○	○	○	○
Mains		Switch Live	E		Pump

Back boiler unit

Heat exchanger material, cast iron
*Permanent live only required if fire front has bulbs

Suitable for use on sealed systems when used on fully pumped systems

Potterton Kingfisher

E	N	L
○	○	○
E	N	Switch Live

Wall mounted

Not suitable for use on sealed systems
Heat exchanger material, cast iron

Potterton Kingfisher 2

E	N	L
○	○	○
E	N	Switch Live

Floor standing

Suitable for use on sealed systems using optional kit
Heat exchanger material, cast iron

Potterton Lynx combination

E	N	L	SWL
○	○	○	○
E	N	Perm Live	Switch Live

Wall mounted

Suitable for sealed systems
Heat exchanger material, copper

Installing and Servicing Domestic Heating Wiring Systems and Control

Potterton Myson Ultra System Boiler   Wall mounted

```
N      L   N   1   2   3   4   5   6   7      ON
O      O   O   O   O   O   O   O   O   O      O
DO NOT Mains                            Switch
USE    240V                             Live
```

Suitable for sealed systems
Heat exchanger material, cast iron

The boiler has the facility to wire external room thermostat, cylinder thermostat and mid-position valve into the control box as follows:

Room stat Common	6	Cylinder stat Common	4	Mid-position valve: Green-yellow	A
Room stat Demand	7	Cylinder stat Demand	ON	Blue	B
Room stat Neutral	2	Cylinder stat Satisfied	5	Grey	C
				Brown or white	D
				Orange	E

Potterton Netaheat MK 1 10/16 and 16/22   Wall mounted

All systems 10/16 and gravity hot water, pumped central heating 16/22

```
E   N   L    F   4   5       6
O   O   O    O   O   O       O
E   N   Perm         Switch
        Live         Live
```

Not suitable for use on sealed systems
Heat exchanger material, cast iron

Fully pumped 16/22

```
E   N   L    F     4     5       6
O   O   O    O     O     O       O
E   N   Perm      Pump  Switch
        Live      Live  Live
```

Link 5–6.

Potterton Netaheat MK 2 10/16 and 16/22   Wall mounted

All systems 10/16 and gravity hot water, pumped central heating 16/22

```
E   N   L    F   4   5       6
O   O   O    O   O   O       O
E   N   Perm         Switch
        Live         Live
```

Not suitable for use on sealed systems
Heat exchanger material, cast iron

Fully pumped 16/22

```
E   N   L    F     4     5       6
O   O   O    O     O     O       O
E   N   Perm      Pump  Switch
        Live      Live  Live
```

Link 5–6.

Boilers – gas

Potterton Netaheat MK 2F 10/16 and 16/22
Wall mounted

All systems 10/16 and gravity hot water, pumped central heating 16/22

```
E   N   L     6    4    5
O   O   O     O    O    O
E   N   Perm       Switch
        Live       Live
```

Not suitable for use on sealed systems
Heat exchanger material, cast iron

Fully pumped 16/22

```
E   N   L     6     4     5
O   O   O     O     O     O
E   N   Perm  Pump  Switch
        Live  Live  Live
```

Link 5–6.

Potterton Netaheat Electronic 6/10, 10/16 and 16/22
Wall mounted

All systems 6/10, 10/16 and gravity hot water, pumped central heating 16/22

```
E   N   L   E   N   6   7   8
O   O   O   O   O   O   O   O
E   N                   Switch
                        Live
```

Suitable for use on sealed systems using optional kit
Heat exchanger material, cast iron

Fully pumped 16/22

```
E   N   L     E   N   6     7      8
O   O   O     O   O   O     O
E   N   Perm          Pump  Switch
        Live          Live  Live
```

Link 7–8.

Potterton Profile (also Netaheat Profile)
Wall mounted

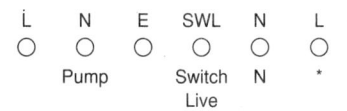

Gravity hot water, pumped central heating

```
L    N    E    SWL    N    L
O    O    O    O      O    O
Pump           Switch N    *
               Live
```

Suitable for use on sealed systems
Heat exchanger material, cast iron

* From room thermostat demand or central heating 'on' if no room thermostat fitted.

Fully pumped

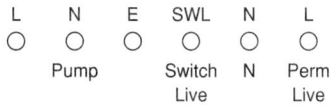

```
L    N    E    SWL    N    L
O    O    O    O      O    O
Pump           Switch N    Perm
               Live        Live
```

135

Installing and Servicing Domestic Heating Wiring Systems and Control

Potterton Profile prima

Gravity hot water, pumped central heating (not fanned flue model)

```
L    N    E   SWL   N    L
O    O    O    O    O    O
          E  Switch  N
             Live
```

Wall mounted

Fanned flue models: fully pumped systems only
Suitable for use on sealed systems
Heat exchanger material, cast iron

Fully pumped

```
L    N    E   SWL   N    L
O    O    O    O    O    O
Pump      Switch  N   Perm
          Live         Live
```

Potterton Puma combination

```
External Timer              Room Stat
10   9    8    7    6    5    4    3    2    1
E    N    L   SWL   E    N    L   SWL   N    L
O    O    O    O    O    O    O    O    O    O
240V supply  Switch  E    N    Room      Mains
To Timer     Live              Stat       240V
```

Wall mounted

Suitable for sealed systems
Heat exchanger material, copper
Internal frost thermostat fitted

Potterton suprima

```
L    N    E   SWL   N    L    E
O    O    O|   O   |O    O    O
Pump      Switch       Mains
          Live
```

Wall mounted

Fully pumped systems only
Suitable for use on sealed systems
using optional kit
Heat exchanger material, cast iron

Potterton Tatler

```
E    N    L
O    O    O
E    N  Switch
        Live
```

Floor standing

Not suitable for use on sealed systems
Heat exchanger material, cast iron

Powermatic SGM 3GB
SGM 4GB

Landis & Gyr RWB2 programmer fitted

```
1    2    3    4    5    6    7    8    9   10   11   12
O    O    O    O    O    O    O    O    O    O    O    O
L    N    N   CH              N    N         N         HW
Mains         ON                                       ON
```

Floor standing

Not suitable for use on sealed systems
Heat exchanger material, cast iron

Terminals 2–3–6–8–10 are linked internally.
Remove links 1–4–12.

External programmer fitted.

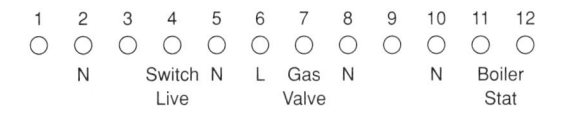

```
 1   2   3   4   5   6   7   8   9   10  11  12
 O   O   O   O   O   O   O   O   O   O   O   O
     N      Switch  N   L   Gas     N       Boiler
            Live            Valve           Stat
```

Control box can be used for connecting pump and room thermostat if required between terminals 5 and 9.

Powermax 195 combination

Floor standing

The bulk of the water in the boiler's thermal store is used for central heating and also for heating the integral heat exchanger coil to supply domestic hot water. Model 185P has integral pump and 185CP has integral pump and programmer. Diagram shows basic model.

```
 L1  N1  E   6   7   8   9   10  11  12  13  14
 O   O   O   O   O   O   O   O   O   O   O   O
             L   N   E   Room    Live  N
             CH Pump    Stat     Out
```

Not suitable for use on sealed systems
Heat exchanger material, copper

An internal programmer should have voltage free contacts. Link programmer live to hot water common. Connect the HW demand to terminal 14. Wire the central heating channel by removing link across 9–10 and wiring CH common and CH demand into 9–10. The programmer live should not be linked to the central heating switch.

Radiant RCM, Comfort, RSF combination

Wall mounted

A 3-core lead is supplied for connection to an adjacent point. External controls such as time clock with voltage free contacts, or room thermostat, should be connected between terminals TA–OR after removing link. Room thermostat neutral should not be connected.

Suitable for sealed systems
Heat exchanger material, copper

Radiant R and RS Comfort heating only

Wall mounted

A 3-core lead is supplied for connection to an adjacent point. External controls such as time clock with voltage free contacts, or room thermostat, should be connected between terminals TA–OR after removing link. Room thermostat neutral should not be connected.

Suitable for sealed systems
Heat exchanger material, copper

Radiant RMA, RMAS combination

Wall mounted

Incorporates 45 litre domestic hot water cylinder

A 3-core lead is supplied for connection to an adjacent point. External controls such as time clock with voltage free contacts, or room thermostat, should be connected between terminals TA–OR after removing link. Room thermostat neutral should not be connected.

Suitable for sealed systems
Heat exchanger material, copper

Installing and Servicing Domestic Heating Wiring Systems and Control

Ravenheat 30B, 40B, 50B, 50S

E	N	L	D	C	N	7
○	○	○	○	○	○	○
E	N	Perm Live	Switch Live		N Pump	L

When wiring external controls remove link D–C.

Wall mounted

Fully pumped systems only
Suitable for use on sealed systems
(50S also incorporates a pressure vessel above the unit)
Heat exchanger material, copper

Ravenheat CF10/20, CF10/25 combination

L	N		C	D
○	○		○	○
Mains 240V			24V	

Wall mounted

Suitable for sealed systems
Heat exchanger material, copper

External controls such as a time clock with voltage free contacts, or room thermostat suitable for use at 240V, should be connected between terminals C–D after removing link.

Ravenheat Combiplus combination

L	N		C	D
○	○		○	○
Mains 240V			External Controls	

Wall mounted

Suitable for sealed systems
Heat exchanger material, copper

External controls such as a time clock with voltage free contacts, or room thermostat suitable for use at 240V, should be connected between terminals C–D after removing link.

Ravenheat Merit

L	N	E	L1	L2	N1	L	N
○	○	○	○	○	○	○	○
	Pump		Switch Live			Mains	

Wall mounted

Fully pumped systems only
Suitable for use on sealed systems
Heat exchanger material, copper

Ravenheat RSF 820/20 combination

L	N		C	D
○	○		○	○
Mains 240V			24V	

Wall mounted

Suitable for sealed systems
Heat exchanger material, copper

External controls such as a time clock with voltage free contacts, or room thermostat suitable for use at 24V, should be connected between terminals C–D after removing link.

Boilers – gas

Rayburn GD 80 cooking appliance/boiler

Floor standing

PL	PN	E	L	N	E	L1	N	CSL	BSL	L2	L3
○	○	○	○	○	○	○	○	○	○	○	○

Pump   Mains   Switch Live

Fully pumped systems only
Suitable for use on sealed systems
Heat exchanger material, copper

If wiring external controls remove link L1–BSL. Pump Live must be wired into terminal PL.

Saunier Duval SD 123C, 123F
SD 223C, 223F

Wall mounted

```
           1   2   3
      N○   ○   ○   ○
L○    E○   ○   ○   ○
  Mains 240V  6   5   4
```

Fully pumped systems only
Suitable for sealed systems
Heat exchanger material, copper

External controls such as a time clock with voltage free contacts, or room thermostat, should be connected between terminals 1–2 after removing link. Room thermostat neutral can be connected to terminal 5.

Saunier Duval SD 135C, 135F
SD 235C, 235F

Wall mounted

```
           1   2   3
      N○   ○   ○   ○
L○    E○   ○   ○   ○
  Mains 240V  6   5   4
```

Fully pumped systems only
Suitable for sealed systems
Heat exchanger material, copper

External controls such as a time clock with voltage free contacts, or room thermostat, should be connected between terminals 1–2 after removing link. Room thermostat neutral can be connected to terminal 5.

Saunier Duval SD 620F combination

Wall mounted

```
           1   2   3
      N○   ○   ○   ○
L○    E○   ○   ○   ○
  Mains 240V  6   5   4
```

Suitable for sealed systems
Heat exchanger material, copper

External controls such as a time clock with voltage free contacts, or room thermostat, should be connected between terminals 2–3 after removing link. Room thermostat neutral can be connected to terminal 5.

Saunier Duval SD 623 combination

```
            1   2   3
      N O   O   O   O
  L O
      E O   O   O   O
  Mains 240V 6   5   4
```

Wall mounted

Suitable for sealed systems
Heat exchanger material, copper

External controls such as a time clock with voltage free contacts, or room thermostat, should be connected between terminals 1–2 after removing link. Room thermostat neutral can be connected to terminal 5.

Saunier Duval SD 625M combination

```
            1   2       3
      N O   O   O  24V  O
  L O
      E O   O   O       O
  Mains 240V 6  5       4
```

Wall mounted

Suitable for sealed systems
Heat exchanger material, copper

External controls such as a time clock with voltage free contacts, or room thermostat suitable for use at 24V, should be connected between terminals 2–3 after removing link. Room thermostat neutral can be connected to terminal 1.

Saunier Duval Master Twin combination

Floor standing

Supplied with mains lead.

Suitable for sealed systems
Heat exchanger material, copper
Integral frost thermostat

External controls such as a time clock with voltage free contacts, or room thermostat, should be connected into 3-pin connector after removing link between terminals 2–3 in plug. Room thermostat neutral should not be connected.

Saunier Duval SB 23 Thelia

Wall mounted

Supplied with mains lead.

Fully pumped systems only
Suitable for sealed systems
Heat exchanger material, copper

External controls such as a time clock with voltage free contacts, or room thermostat, should be connected into 5-pin connector between top two terminals after removing link. Room thermostat neutral should not be connected. A cylinder thermostat can be connected across the bottom two terminals.

Boilers – gas

Saunier Duval 23, 23E Thelia combination

Wall mounted

Supplied with mains lead.

Suitable for sealed systems
Heat exchanger material, copper

External controls such as a time clock with voltage free contacts, or room thermostat, should be connected into 3-pin connector after removing link between 2–3 in plug. Room thermostat neutral should not be connected.

Saunier Duval 30E Thelia combination

Wall mounted

```
              1   2   3
       N O    O   O   O
L O    E O    O   O   O
Mains 240V    6   5   4
```

Suitable for sealed systems
Heat exchanger material, copper

External controls such as a time clock with voltage free contacts, or room thermostat, should be connected between terminals 2–3 after removing link. Room thermostat neutral can be connected to terminal 5.

Saunier Duval Thelia Twin combination

Wall mounted

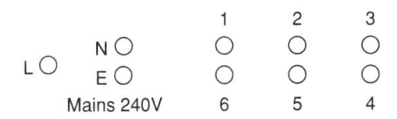

```
              1   2   3
       N O    O   O   O
L O    E O    O   O   O
Mains 240V    6   5   4
```

Suitable for sealed systems
Heat exchanger material, copper
Integral frost thermostat

External controls such as a time clock with voltage free contacts, or room thermostat, should be connected between terminals 2–3 after removing link. Room thermostat neutral must not be connected.

Saunier Duval 223 Themis combination

Wall mounted

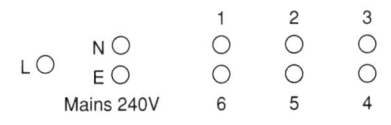

```
              1   2   3
       N O    O   O   O
L O    E O    O   O   O
Mains 240V    6   5   4
```

Suitable for sealed systems
Heat exchanger material, copper

For use without time switch or room thermostat, link together terminals 1, 2 and 3. With this wiring the control thermostat controls the operation of the burner – the pump runs continuously. To stop the pump working this way, disconnect the wiring loom connectors and insulate them from the earth of the boiler. With this layout the control thermostat controls the burner. The pump will stop when the burner goes out. For use with time switch and room thermostat, connect time switch voltage free contacts and room thermostat in series between terminals 1 and 2 and link together terminals 2 and 3. The neutral of the room thermostat must be connected to terminal 5.

Saunier Duval System 400

Electrical connection is via a 3-pin plug

Wall mounted

Fully pumped systems only
Suitable for use on sealed systems
Heat exchanger material, copper

Sime Super 90 combination

Connection via internal 5 pin plug

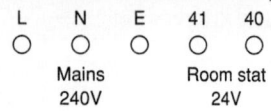

```
L     N     E     41    40
O     O     O     O     O
   Mains        Room stat
   240V           24V
```

Wall mounted

Suitable for sealed systems
Heat exchanger material, copper
Frost thermostat and time clock
integral to boiler.

Sinclair 40, 50

```
E    L    N    NP   LP   EP   1    2    3   4
O    O    O    O    O    O    O    O    O   O
E    L    N         Pump      L   Switch  Internal
     Mains                         Live    Wiring
```

Wall mounted

Fully pumped systems only
Not suitable for use on sealed systems
Heat exchanger material, copper

Thermomatic RSM 15, 20, 25 combination

```
14    13    12    11    E    N    L
O     O     O     O     O    O    O
                        Mains
```

Wall mounted

Suitable for sealed systems
Heat exchanger material, copper

External controls such as a time clock with voltage free contacts, or room thermostat, should be connected between terminals 11–12 after removing link. Terminals 13–14 should remain linked.

Trianco Homeflame

```
  L     N     E
  O     O     O
Switch  N     E
 Live
```

Back boiler unit

Not suitable for use on sealed systems
Heat exchanger material, cast iron

Trianco Triancogas F

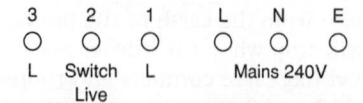

```
 3     2     1     L     N     E
 O     O     O     O     O     O
 L   Switch  L      Mains 240V
      Live
```

Wall mounted

Fully pumped systems only
Suitable for use on sealed systems
Heat exchanger material, copper

Trianco Triancogas RS

```
E    N    L
O    O    O
E    N    Switch
          Live
```

Wall mounted

Fully pumped systems only
Suitable for use on sealed systems
Heat exchanger material, copper

Trianco Tristar

```
E    N    L     EP   NP   LP    1    2
O    O    O     O    O    O     O    O
E    N    L     E    N    L     Switch
  Mains           Pump           Live
```

Terminal 1 can be used as live to external controls

Wall mounted

Fully pumped systems only
Suitable for use on sealed systems
Heat exchanger material, copper

Trisave FS 12–18, FS 18–24 condensing

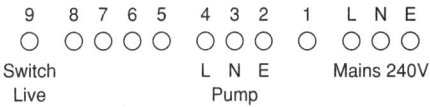

```
9    8  7  6  5    4 3 2    1    L  N  E
O    O  O  O  O    O O O    O    O  O  O
Switch             L N E         Mains 240V
Live               Pump
```

Remove Link 1–9.

Floor standing

Fully pumped systems only
Suitable for use on sealed systems
using optional kit
Heat exchanger material, aluminium

Trisave FS 60, 80, 80C condensing

A 4-core lead is supplied which should
be connected as follows:
Grn yllw   Earth
Blue   Neutral
Black   Perm live
Brown   Switch live

Floor standing

Fully pumped systems only
Suitable for use on sealed systems
using optional kit
Heat exchanger material, aluminium
The pump is connected into the boiler
via 3-pin plug supplied.

Trisave Turbo 22, 30, 45, 60 condensing

A 4-core lead is supplied which should
be connected as follows:
Grn yllw   Earth
Blue   Neutral
Black   Perm live
Brown   Switch live

Wall mounted

Fully pumped systems only
Suitable for use on sealed systems
using optional kit
Heat exchanger material, aluminium
The pump is connected into the boiler
via 3-pin plug supplied.

Vaillant, general information

Model codes:

VC = Vaillant Central (heating)
VCW = Vaillant Central (heating) Water, e.g. Combi
GB = Great Britain – following models may include GB but wiring is the same.
First two numbers denote kilowatt output and third number denotes flue type, i.e.
0 = Conventional, 1= Balanced, 2 = Fanned
Following letters denote thus: E = Electronic, H = Natural gas, B = bottled gas.
Therefore a VCC 242E is a 24 kilowat fanned flue electronic combination boiler.

Vaillant VC 10, 15 TW3, 20 TW3   Wall mounted

```
6   5   4   3   2R  1MP
O   O   O   O   O   O
        External  L   N
        Switches  Mains 240V
```

Fully pumped systems only
Suitable for sealed systems
Heat exchanger material, copper

External controls such as a time clock with voltage free contacts, or room thermostat, should be connected between terminals 3–4 after removing link.

Vaillant VC 10-W, 15-W, 20-W   Wall mounted

```
6   5   4   3   2R  1MP
O   O   O   O   O   O
        External  L   N
        Controls  Mains 240V
```

Fully pumped systems only
Suitable for sealed systems
Heat exchanger material, copper

External controls such as a time clock with voltage free contacts, or room thermostat, should be connected between terminals 3–4 after removing link.

Vaillant VC 110, 180, 240 combination   Wall mounted

```
        N   L
N   L   1   2   3   4   5
O   O   O   O   O   O   O
        Mains
        240V
```

Fully pumped systems only
Suitable for sealed systems
Heat exchanger material, copper

External controls such as a time clock with voltage free contacts, or room thermostat, should be connected between terminals 3–4 after removing link. Connect room thermostat neutral to terminal 5. Do not use terminals 7–12.

Boilers – gas

Vaillant VC 110 H, 180 H, 240 H

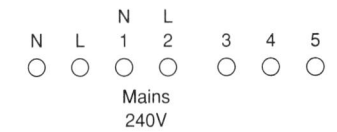

```
      N  L
N  L  1  2   3  4  5
O  O  O  O   O  O  O
      Mains
      240V
```

Wall mounted

Fully pumped systems only
Suitable for sealed systems
Heat exchanger material, copper

External controls such as a time clock, or room thermostat, should be connected between terminals 3–4. Connect room thermostat neutral to terminal 5. Do not use terminals 7–12.

Vaillant VC 112 EH, 142 EH

```
      N  L
N  L  1  2   3  4  5
O  O  O  O   O  O  O
      Mains
      240V
```

Wall mounted

Fully pumped systems only
Suitable for sealed systems
Heat exchanger material, copper

External controls such as a time clock, or room thermostat, should be connected between terminals 3–4. Connect room thermostat neutral to terminal 5. Do not use terminals 7–12.

Vaillant VC 112 E – RSF VC 142 E, 182 E, 242 E, 282 E

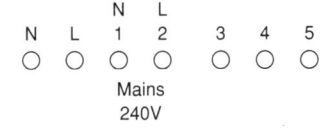

```
      N  L
N  L  1  2   3  4  5
O  O  O  O   O  O  O
      Mains
      240V
```

Wall mounted

Fully pumped systems only
Suitable for sealed systems
Heat exchanger material, copper

External controls such as a time clock, or room thermostat, should be connected between terminals 3–4. Connect room thermostat neutral to terminal 5. Do not use terminals 7–12.

Vaillant VC 221 H

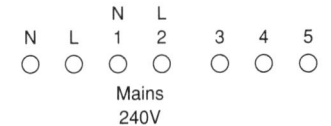

```
      N  L
N  L  1  2   3  4  5
O  O  O  O   O  O  O
      Mains
      240V
```

Wall mounted

Fully pumped systems only
Suitable for sealed systems
Heat exchanger material, copper

External controls such as a time clock, or room thermostat, should be connected between terminals 3–4. Connect room thermostat neutral to terminal 5. Do not use terminals 7–12.

Vaillant VC Sine 18 W

```
6   5   4   3   2R  1MP
O   O   O   O   O   O
        External  L   N
        Controls  Mains 240V
```

Wall mounted

Fully pumped systems only
Suitable for sealed systems
Heat exchanger material, copper

External controls such as a time clock with voltage free contacts, or room thermostat, should be connected between terminals 3–4 after removing link.

Vaillant VCW 20 T3W, 25 T3W combination

```
6   5   4   3   2R  1MP
O   O   O   O   O   O
        External  L   N
        Controls  Mains 240V
```

Wall mounted

Suitable for sealed systems
Heat exchanger material, copper

External controls such as a time clock with voltage free contacts, or room thermostat, should be connected between terminals 3–4 after removing link.

Vaillant VCW 182 E, 242 E combination

```
        N   L
N   L   1   2   3   4   5
O   O   O   O   O   O   O
        Mains
        240V
```

Wall mounted

Suitable for sealed systems
Heat exchanger material, copper

External controls such as a time clock, or room thermostat, should be connected between terminals 3–4. Connect room thermostat neutral to terminal 5. Do not use terminals 7–12.

Vaillant VC 221 combination

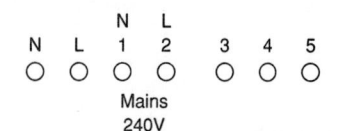

```
        N   L
N   L   1   2   3   4   5
O   O   O   O   O   O   O
        Mains
        240V
```

Wall mounted

Fully pumped systems only
Suitable for sealed systems
Heat exchanger material, copper

External controls such as a time clock, or room thermostat, should be connected between terminals 3–4. Connect room thermostat neutral to terminal 5. Do not use terminals 7–12.

Vaillant VCW 240 H, 280 H combination

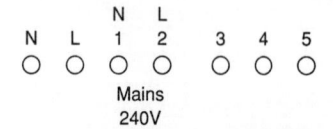

```
        N   L
N   L   1   2   3   4   5
O   O   O   O   O   O   O
        Mains
        240V
```

Wall mounted

Suitable for sealed systems
Heat exchanger material, copper

External controls such as a time clock, or room thermostat, should be connected between terminal 3–4. Connect room thermostat neutral to terminal 5. Do not use terminals 7–12.

Boilers – gas

Vaillant VWC Sine 18 T3W combination

Wall mounted

```
 6    5    4    3   2R  1MP
 ○    ○    ○    ○    ○   ○
           External   L   N
           Controls  Mains 240V
```

Suitable for sealed systems
Heat exchanger material, copper

External controls such as a time clock with voltage free contacts, or room thermostat, should be connected between terminals 3–4 after removing link.

Vaillant VCW 242 EH, 282 EH combination

Wall mounted

```
          N   L
 N   L    1   2    3   4   5
 ○   ○    ○   ○    ○   ○   ○
          Mains
          240V
```

Suitable for sealed systems
Heat exchanger material, copper

External controls such as a time clock, or room thermostat, should be connected between terminals 3–4. Connect room thermostat neutral to terminal 5. Do not use terminals 7–12.

Vaillant VK-E, VKS-E

Floor standing

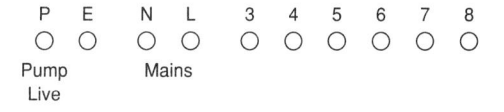

```
 P   E    N   L    3   4   5   6   7   8
 ○   ○    ○   ○    ○   ○   ○   ○   ○   ○
 Pump     Mains
 Live
```

Fully pumped systems only
Suitable for use on sealed systems
Heat exchanger material, cast iron

External controls such as a time clock with voltage free contacts, or room thermostat, should be connected between terminals 3–4 after removing link.

Vaillant VU 186 EH
VU 226 EH condensing, system

Wall mounted

```
 E   N   L    3   4   5    7   8   9
 ○   ○   ○    ○   ○   ○    ○   ○   ○
 E   N   L              DO NOT
    Mains                USE
```

Fully pumped systems only
Suitable for use on sealed systems
Heat exchanger material, copper

External controls such as a time clock with voltage free contacts, or room thermostat, should be connected between terminals 3–4 after removing link. When used in conjunction with a 3-port or 2 × 2 port motorized valve, the switched live can be connected into terminal 4.

Installing and Servicing Domestic Heating Wiring Systems and Control

Vaillant VUW 236 EH
VUW 286 EH
condensing, combination

Wall mounted

```
E  N  L     3  4  5      7  8  9
O  O  O     O  O  O      O  O  O
E  N  L                  DO NOT
  Mains                    USE
```

Suitable for sealed systems
Heat exchanger material, copper

External controls such as a time clock with voltage free contacts, or room thermostat, should be connected between terminals 3–4 after removing link.

Vokera 12–48 RS Mynute
12–48 RSE Mynute

Wall mounted

```
N  L   1  2  3  4
O  O   O  O  O  O
Mains 240V  External
            Controls
```

Suitable for sealed systems
Heat exchanger material, copper

Connect external controls such as a time clock, or room thermostat between terminals 2–3 after removing link. If wiring into a conventional fully pumped system, e.g. Y Plan, the switch live should be connected to terminal 3 after removing link.

Vokera 18–72 DMCF
21–84 DMCF combination

Wall mounted

```
N  L   3  4  5
O  O   O  O  O
Mains  External Controls
240V   240V
```

Suitable for sealed systems
Heat exchanger material, copper

Connect external controls such as a time clock, or room thermostat, between terminals 3–4 after removing link. Connect room thermostat neutral into terminal 5.

Vokera 20–80 flowmatic combination

Wall mounted

```
N  L   1  2  3   N
O  O   O  O  O   O
Mains  External  N
240V   Controls
```

Suitable for sealed systems
Heat exchanger material, copper

Connect external controls such as a time clock, or room thermostat, between terminals 1–3.

Boilers – gas

Vokera 20–80 RS turbo combination

```
N   L   3   4   5
O   O   O   O   O
Mains   External Controls
240V    240V
```

Wall mounted

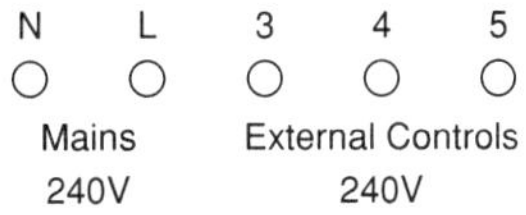

Suitable for sealed systems
Heat exchanger material, copper

Connect external controls such as a time clock, or room thermostat, between terminals 3–4 after removing link. Connect room thermostat neutral into terminal 5.

Vokera 21–84 Turbo combination

```
E   N   L   4   5
O   O   O   O   O
Mains 240V    External
              Controls
```

Wall mounted

Suitable for sealed systems
Heat exchanger material, copper

Connect external controls such as a time clock with voltage free contacts, or room thermostat, between terminals 4–5 after removing link.

Vokera 21–84 DC turbo combination

```
N   L   3   4   5
O   O   O   O   O
Mains   External Controls
240V    240V
```

Wall mounted

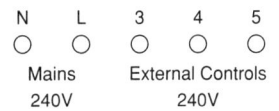

Suitable for sealed systems
Heat exchanger material, copper

Connect external controls such as a time clock, or room thermostat, between terminals 3–4 after removing link. Connect room thermostat neutral into terminal 5.

Vokera 20–80 Flowmatic combination

```
N   L   1   2   3   N
O   O   O   O   O   O
Mains   External     N
240V    Controls
```

Wall mounted

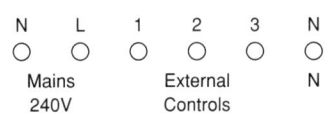

Suitable for sealed systems
Heat exchanger material, copper

Connect external controls such as a time clock, or room thermostat, between terminals 1–3.

Warm World HE30–HE70 condensing

```
L L   N N N   1   2   3   4   5   6
OO    OOO     O   O   O   O   O   O
L     N       Pump Switch HW  HW  CH
              Live Live   Off On  On
                          Integral
                          Programmer
                          if fitted
```

Wall mounted

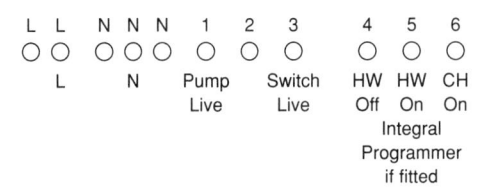

Fully pumped systems only
Suitable for use on sealed systems
Heat exchanger material, aluminium and copper

Wickes 40, 50

```
 4  3    2    1   EP  LP  NP    N   L   E
 O  O    O    O   O   O   O     O   O   O
Internal Switch  L   E   L   N      N   L   E
Wiring   Live        Pump              Mains
```

Wall mounted

Fully pumped systems only
Not suitable for use on sealed systems
Heat exchanger material, copper

Wickes 40, 50 MK2

```
 4  3    2    1   EP  LP  NP    N   L   E
 O  O    O    O   O   O   O     O   O   O
Internal Switch  L   E   L   N      N   L   E
Wiring   Live        Pump              Mains
```

Wall mounted

Fully pumped systems only
Suitable for use on sealed systems
Heat exchanger material, copper

Wickes 42, 53

```
 E  L  N    NP  LP  EP    1     2     3   4
 O  O  O    O   O   O     O     O     O   O
 E  L  N        Pump      L   Switch  Internal
   Mains                       Live    Wiring
```

Wall mounted

Fully pumped systems only
Not suitable for use on sealed systems
Heat exchanger material, copper

Wickes 45F, 65F

```
12 11 10   9 8    7    6    5 4 3 2 1
 O  O  O   O O    O    O    O O O O O
 L  N  E   L N    L   Switch  Internal Wiring
   Mains   Pump       Live
```

Wall mounted

Fully pumped systems only
Suitable for use on sealed systems
Heat exchanger material, copper

Terminals 7 and 12 are linked

Wickes Combi

```
 L   N    1    2     3    4    5     6    E
 O   O    O    O     O    O    O     O    O
 L   N    N    L    Out   In  Out    In   E
Mains    Timer      Timer     Room Stat
         Motor     Contacts    Contacts
```

Wall mounted

Suitable for sealed systems
Heat exchanger material, copper

If no timer or room stat Link 3–4 or 5–6 as appropriate

Wickes Combi 30/90 combination

Connection via internal plug

```
 L     N     E    41    40
 O     O     O    O     O
      Mains      Room stat 24V
```

Wall mounted

Suitable for sealed systems
Heat exchanger material, copper

If no room thermostat link 40–41. Time clock integral to boiler.

Boilers – gas

Worcester 9.24 BF, OF electronic, combination

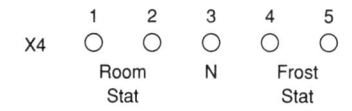

Wall mounted

```
         1    2    3    4    5
X4       O    O    O    O    O
       Room       N   Frost
       Stat           Stat
```

Suitable for sealed systems
Heat exchanger material, copper

Worcester 9.24 BF, OF, MK 1.5 combination

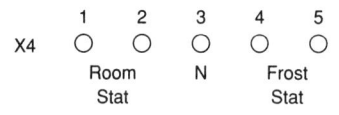

Wall mounted

```
         1    2    3    4    5
X4       O    O    O    O    O
       Room       N   Frost
       Stat           Stat
```

Suitable for sealed systems
Heat exchanger material, copper

Worcester 9.24 RSF
9.24 RSF 'E'
electronic, combination

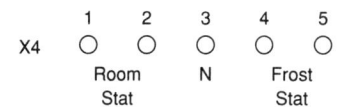

Wall mounted

```
         1    2    3    4    5
X4       O    O    O    O    O
       Room       N   Frost
       Stat           Stat
```

Suitable for sealed systems
Heat exchanger material, copper

Worcester 24 CDi combination

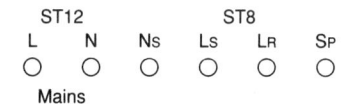

Wall mounted

```
     ST12           ST8
  L   N    Ns   Ls   Lr   Sp
  O   O    O    O    O    O
 Mains
```

Suitable for sealed systems
Heat exchanger material, copper

External controls such as a time clock with voltage free contacts, or room thermostat, should be connected between terminals LS–LR after removing link. Room thermostat neutral can be connected to NS. If system uses wire-free remote Digistat room thermostat – see ACL-Drayton Digistat RF1. If external frost protection is required also connect across LS–LR in addition to other wiring.

Worcester 240, BF, OF

Wall mounted

```
          1      2    3    4    5
X4        O      O    O    O    O
       Room Stat      N   Frost
       or Timer           Stat
```

Suitable for sealed systems
Heat exchanger material, copper

External controls such as a time clock with voltage free contacts, or room thermostat suitable for use at 240V, should be connected as shown.

151

Installing and Servicing Domestic Heating Wiring Systems and Control

Worcester 240, BF, OF RSF

```
         1    2    3    4    5
X4       O    O    O    O    O
      Room Stat   N    Frost
      or Timer         Stat
```

Wall mounted

Suitable for sealed systems
Heat exchanger material, copper

External controls such as a time clock with voltage free contacts, or room thermostat suitable or use at 240V, should be connected as shown.

Worcester 280, RSF

```
         5    4    3    2    1
X5       O    O    O    O    O
         N  Room Stat  Frost
            or Timer   Stat
```

Wall mounted

Suitable for sealed systems
Heat exchanger material, copper

External controls such as a time clock with voltage free contacts, or room thermostat suitable for use at 240V, should be connected as shown.

Worcester 350 combination

```
              N    N L
X4            O    O O    O    O O    O O
                   Mains       Frost  Room
                   240V        Stat   Stat

X14   From CH on      O    External
      From HW on      O    Programmer
      Live to Prof.   O    Connections
      Neutral to prog. O
```

Wall mounted

Suitable for sealed systems
Heat exchanger material, copper

Worcester Delglo 2 combination

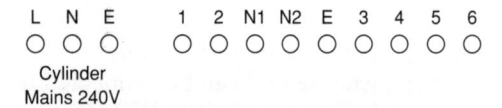

```
   L  N  E      1  2  N1 N2 E  3  4  5  6
   O  O  O      O  O  O  O  O  O  O  O  O
   Cylinder
   Mains 240V
```

Floor standing

Suitable for use on sealed systems
Heat exchanger material, copper

External controls such as a time clock with voltage free terminals, or room thermostat, should be connected between terminals 2–3 after removing link. A frost thermostat should be of the double pole type wired common to terminal 6 and 2 pole connections to 1 and 2.

Worcester Delglo 3 combination

```
   L  N  E      1  2  N1 N2 E  3  4  5  6
   O  O  O      O  O  O  O  O  O  O  O  O
   Cylinder
   Mains 240V
```

Floor standing

Suitable for use on sealed systems
Heat exchanger material, copper

External controls such as a time clock with voltage free terminals, or room thermostat, should be connected between terminals 2–3 after removing link. A frost thermostat should be of the double pole type wired common to terminal 6 and 2 pole connections to 1 and 2.

Boilers – gas

Worcester Heatslave 9.24 BF, OF, MK 1 combination

Floor standing

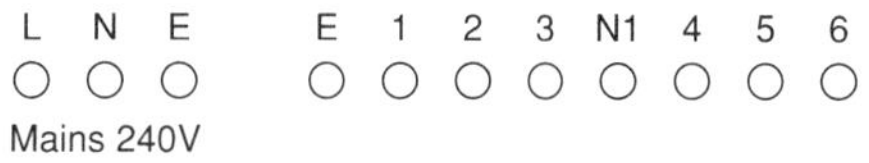

L N E   E 1 2 3 N1 4 5 6
○ ○ ○   ○ ○ ○ ○ ○ ○ ○ ○
Mains 240V

Suitable for use on sealed systems
Heat exchanger material, copper

External controls such as a time clock with voltage free terminals, or room thermostat, should be connected between terminals 2–3 after removing link. Wire frost thermostat across terminals 4–5.

Worcester Heatslave 9.24 RSF combination

Wall mounted

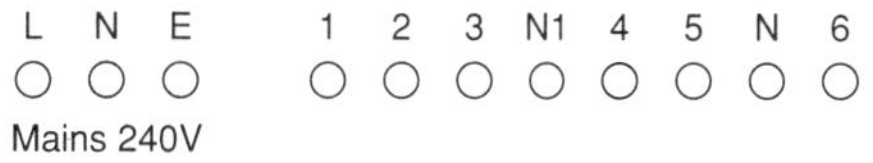

L N E   1 2 3 N1 4 5 N 6
○ ○ ○   ○ ○ ○ ○ ○ ○ ○ ○
Mains 240V

Suitable for use on sealed systems
Heat exchanger material, copper

External controls such as a time clock with voltage free terminals, or room thermostat, should be connected between terminals 2–3 after removing link. Wire frost thermostat across main live and terminal 3.

Worcester Heatslave High Flow combination

Floor standing

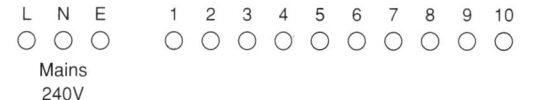

L N E   1 2 3 4 5 6 7 8 9 10
○ ○ ○   ○ ○ ○ ○ ○ ○ ○ ○ ○ ○
Mains 240V

Suitable for use on sealed systems
Heat exchanger material, copper

External controls such as a time clock with voltage free terminals, or room thermostat, should be connected between terminals 2–3 after removing link. A frost thermostat should be of the double pole type and connected common to terminal L and 2 pole connections to 1 and 3.

Worcester Heatslave Junior 12 combination

Wall mounted

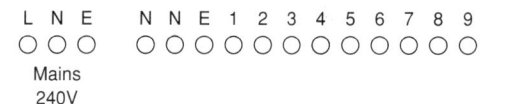

L N E   N N E 1 2 3 4 5 6 7 8 9
○ ○ ○   ○ ○ ○ ○ ○ ○ ○ ○ ○ ○ ○ ○
Mains 240V

Suitable for use on sealed systems
Heat exchanger material, copper

External controls such as a time clock with voltage free terminals, or room thermostat, should be connected between terminals 2–4 after removing link. A frost thermostat should be of the double pole type and wired common to terminal 3 and 2 pole connections to 1 and 2.

Installing and Servicing Domestic Heating Wiring Systems and Control

Worcester Heatslave Senior 6 combination

```
L  N  E    N  N  E  1  2  3  4  5  6  7  8  9
O  O  O    O  O  O  O  O  O  O  O  O  O  O  O
 Mains
 240V
```

Floor standing

Suitable for use on sealed systems
Heat exchanger material, copper

External controls such as a time clock with voltage free terminals, or room thermostat, should be connected between terminals 2–4 after removing link. A frost thermostat should be of the double pole type and wired common to terminal 3 and 2 pole connections to 1 and 2.

Worcester High Flow 3.5 BF, OF, 4.5 BF, OF, 5.3 BF, OF combination

```
L  N  E      1  2  3  4  5  6  7  8  9  10
O  O  O      O  O  O  O  O  O  O  O  O  O
 Mains
 240V
```

Floor standing

SS versions only, suitable for use on sealed systems
Heat exchanger material, copper

External controls such as a time clock with voltage free terminals, or room thermostat, should be connected between terminals 2–3 after removing link. A frost thermostat should be of the double pole type and connected common to terminal L and 2 pole connections to 1 and 3.

Worcester Senior 15, 20, 30 combination

```
L  N  E    N  N  E  1  2  3  4  5  6  7  8  9
O  O  O    O  O  O  O  O  O  O  O  O  O  O  O
 Mains
 240V
```

Floor standing

Not suitable for use on sealed systems
Heat exchanger material, copper

External controls such as a time clock with voltage free terminals, or room thermostat, should be connected between terminals 2–4 after removing link. A frost thermostat should be of the double pole type and wired common to terminal 3 and 2 pole connections to 1 and 2.

Yorkpark Microstar 20 condensing

```
TH   TH        N    N    L
O    O         O    O    O
                    N   Switch
                        Live
```

Wall mounted

Fully pumped systems only
Suitable for sealed systems
Heat exchanger material, aluminium

The above connections apply to a fully pumped system such as a Y plan. However, if only a room thermostat and/or time clock is to be used, then a permanent live should be wired to L and N, and the control wiring connected across TH–TH after removing link.

Boilers – gas

Yorkpark Microstar 20 condensing, combination

Wall mounted

```
TH  TH      N   N   L
O   O       O   O   O
            N   Switch
                Live
```

Suitable for sealed systems
Heat exchanger material, aluminium

The above connections apply to a fully pumped system such as a Y plan. However, if only a room thermostat and/or time clock is to be used, then a permanent live should be wired to L and N and the control wiring connected across TH–TH after removing link.

Yorkpark Microstar MC24G condensing

Floor standing

```
16 15    14 13    9 8     7 6      5 4 3    2 1
O  O     O  O     O O     O O      O O O    O O
External Timer    Room    Temp     E N L    Cylinder
Sensor   Contacts Stat    Limit    Diverter Stat
supplied Voltage          Stat     Valve
         free
```

Fully pumped systems only
Suitable for sealed systems
Heat exchanger material, steel

Terminals 1, 2, 6, 7, 8, 9, 13, 14, 15, 16 are low voltage. A 240V mains should be connected into the separate terminal block.

Yorkpark Microstar MZ22C condensing

Wall mounted

```
TH  TH      N   N   L
O   O       O   O   O
            N   Switch
                Live
```

Fully pumped systems only
Suitable for sealed systems
Heat exchanger material, aluminium

The above connections apply to a fully pumped system such as a Y plan. However, if only a room thermostat and/or time clock is to be used, then a permanent live should be wired to L and N, and the control wiring connected across TH–TH after removing link.

Yorkpark Microstar MZ22S condensing, combination

Wall mounted

```
TH  TH      N   N   L
O   O       O   O   O
            N   l
```

Suitable for sealed systems
Heat exchanger material, aluminium

External controls such as a time clock with voltage free terminals, or room thermostat, should be connected between terminals TH–TH after removing link.

9

Boiler wiring – oil

Boulter Camray 2/3

```
L       N       E
O       O       O
Control N       E
Live
```

Camray 2 de-luxe has integral programmer

Floor mounted

Suitable for use on sealed systems
using optional kit
Heat exchanger material, mild steel

Boulter Camray 15/21

```
3       2       1       E
O       O       O
Perm    N    Control
Live         Live
```

Can be mounted externally with GRP case.

Wall mounted

Suitable for use on sealed systems
Heat exchanger material, mild steel
Integral frost thermostat

Boulter Camray Combi combination

Mains via 3-pin plug. External controls, such as a time clock with voltage free contacts or room thermostat should be connected across terminals 3–5.

Floor mounted

Suitable for use on sealed systems
using optional kit
Heat exchanger material, mild steel

Boulter Camray Compact

Connect to external controls via
3-pin plug supplied

Wall mounted

Fully pumped systems only
Suitable for use on sealed systems
Heat exchanger material, mild steel

Boulter Camray L60

```
Perm  Control
Live  Live    N     E
O     O       O     O
Laux  L       N     E
```

Not suitable for use on sealed systems
Heat exchanger material, mild steel

Boiler wiring – oil

Boulter Camray Pathfinder C, PJ

```
L      N    E
O      O    O
Control  N    E
Live
```

Floor mounted

Fully pumped system only
Suitable for use on sealed systems
Heat exchanger material, cast iron

Gemini Triple Pass

```
L1    N2   E3   E4   N5   L6
O     O    O    O    O    O
Control N    E    E    N    L
Live            Burner supply
```

Floor mounted

Suitable for use on sealed systems
Heat exchanger material, steel

Grant Combi 70 and 90 combination

Floor mounted

External controls such as time clock with voltage free contacts or room thermostat should be connected into time clock socket after removing red link.

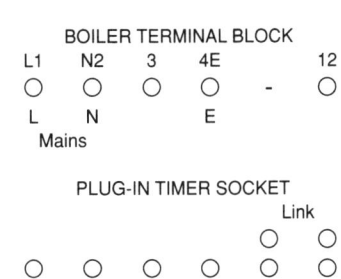

```
      BOILER TERMINAL BLOCK
L1   N2    3    4E          12
O    O     O    O     -     O
L    N          E
Mains

      PLUG-IN TIMER SOCKET
                      Link
                      O    O
O    O    O    O    O    O
```

Suitable for use on sealed systems
Heat exchanger material, steel

Grant Euroflame Utility

```
L1    N2   N3   E4   N5   L6
O     O    O    O    O    O
Control N    E    E    N    L
Live            Burner supply
```

Floor mounted

Suitable for use on sealed systems
Heat exchanger material, steel

Grant Multi Pass

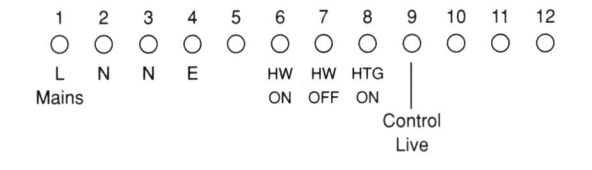

```
1    2    3    4    5    6    7    8    9   10   11   12
O    O    O    O    O    O    O    O    O    O    O    O
L    N    N    E         HW   HW   HTG
Mains                    ON   OFF  ON
                                   Control
                                   Live
```

Floor mounted

Suitable for use on sealed systems
Heat exchanger material, steel

HRM Wallstar

```
O    O    O    O
SW   N    P-L  E
Live
```

Wall mounted

Integral frost thermostat
Suitable for use on sealed systems
Heat exchanger material, mild steel

Installing and Servicing Domestic Heating Wiring Systems and Control

HRM Starflow   **Floor mounted**

```
SW-L  ○
  N   ○
  E   ○
```

Suitable for use on sealed systems
Heat exchanger material, mild steel

Perrymatic Jetstreme MK3   **Floor mounted**

```
CONTROL BOX          TERMINAL BLOCK
1   2   3   4        5    6   7   8   9
○   ○   ○   ○        ○    ○   ○   ○   ○
        N                 Control
                          Live*
```

Suitable for use on sealed systems
Heat exchanger material, steel

*via separate terminal block under burner cover

Thorn Panda   **Floor mounted**

```
E   N   L   1    2    3    L   N   4
○   ○   ○   ○    ○    ○    ○   ○   ○
  *Mains 240V   HW   CH   Control  Pump
                ON*  ON*  Live     Live
```

Not suitable for use on sealed systems
Heat exchanger material, steel

*If integral programmer fitted.
If used on fully pumped system remove links L–1–3, L–2. Link 3–pump live

Trianco Centrajet 13/17   **Wall mounted**

```
E   N   L
○   ○   ○   ○   ○
E   N   Control
        Live
```

Not suitable for use on sealed systems
Heat exchanger material, mild steel

Trianco Centramatic 13/17   **Floor mounted**

Without integral programmer
```
   1        2       3    4
   ○        ○       ○    ○
Control   Perm      N
Live      Live
```

Not suitable for use on sealed systems
Heat exchanger material, mild steel

Remove Link 1–2.

Trianco Centramatic 40, 55, 80   **Floor mounted**

2 way connector on 40 and 55
```
   1     2     3    4
   ○     ○     ○    ○
Control
Live
```

Not suitable for use on sealed systems
Heat exchanger material, mild steel

Remove Link 1–2

A 240V connection should be made to L and N on the Redfyre control box.

Trianco TRO MK1

```
1    2    3    4    5
O    O    O    O    O
Control
Live
```

Floor mounted

Not suitable for use on sealed systems
Heat exchanger material, mild steel

Trianco TRO MK2 and 3

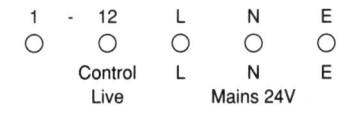

```
1  -  12    L    N    E
O     O     O    O    O
      Control L    N    E
      Live       Mains 24V
```

Remove Link 2–12

Floor mounted

Not suitable for use on sealed systems
Heat exchanger material, mild steel

Trianco TRO 80 combination

```
O    O    O    O    O    O
E    N    L    Control
Mains          Live
```

Wall mounted

Suitable for use on sealed systems
Heat exchanger material, mild steel

Trianco TRO 110 combination

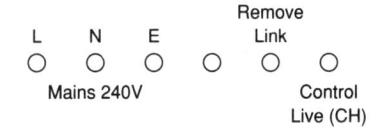

```
                    Remove
L    N    E         Link
O    O    O    O    O    O
Mains 240V          Control
                    Live (CH)
```

Floor mounted

Suitable for use on sealed systems
Heat exchanger material, mild steel

Trianco TRO BF range

```
L      N    E    SL
O      O    O    O
Perm   N    E    Control
Live             Live
```

Remove Link L–2 in 6 way plug-in terminal block.

Floor mounted

Not suitable for use on sealed systems
Heat exchanger material, mild steel

Trianco TSB system boiler

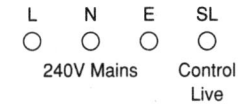

```
L          N    E    SL
O          O    O    O
240V Mains      Control
                Live
```

Remove Link L–1 in 6 way plug-in terminal block.

Wall mounted

Suitable for sealed systems
Heat exchanger material, mild steel

Boiler wiring – oil

Installing and Servicing Domestic Heating Wiring Systems and Control

Trianco Eurostar Combi combination   Floor mounted

```
N   N   E   N   L
O   O   O   O
        240V Mains
```

Suitable for use on sealed systems
Heat exchanger material, mild steel

External controls such as time clock with voltage free contacts or room thermostat should be connected between 3–6 after removing link.

Trianco Eurostar Standard   Floor mounted

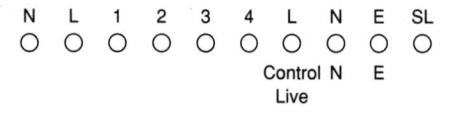

```
N   L   1   2   3   4   L   N   E   SL
O   O   O   O   O   O   O   O   O   O
                    Control N   E
                    Live
```

Suitable for use on sealed systems
Heat exchanger material, mild steel

Trianco Eurostar System   Floor mounted

```
N   L   1   2   3   4   L   N   E   SL
O   O   O   O   O   O   O   O   O   O
                    Control N   E
                    Live
```

Fully pumped systems only
Suitable for use on sealed systems
Heat exchanger material, mild steel

Trianco Eurostar Utility   Floor mounted

```
N   L   1   2   3   4   L   N   E   SL
O   O   O   O   O   O   O   O   O
                    Control N   E
                    Live
```

Suitable for use on sealed systems
Heat exchanger material, mild steel

Trianco Eurostar WM system boiler   Wall mounted

```
L   N   E
O   O   O   O
Control N   E
Live
```

Fully pumped systems only
Suitable for use on sealed systems
Heat exchanger material, mild steel

Worcester Danesmoor DF   Floor mounted

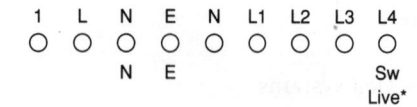

```
1   L   N   E   L1  L2  L3  L4
O   O   O   O   O   O   O   O
        N   E           Sw
                        Live*
```

Suitable for use on sealed systems
using optional kit
Heat exchanger material, mild steel

Remove links L–L2 and L3–L4.

Boiler wiring – oil

Worcester Danesmoor PJ MK1 combination

L	N	E	1	2	3
○	○	○	○	○	○

Mains 240V   Switch Live

Remove link L–3.

Floor mounted

Suitable for use on sealed systems using optional kit
Heat exchanger material, mild steel

Worcester Danesmoore PJ MK2

With external controls

L	N	E	N	1	2	3	4	5	6
○	○	○	○	○	○	○	○	○	○

240V Mains   Switch Live

Remove link plug from programmer terminal strip.

Floor mounted

Suitable for use on sealed systems using optional kit
Heat exchanger material, mild steel

Worcester Danesmoore SLPJ

With external controls

L	N	E	N	1	2	3	4	5	6
○	○	○	○	○	○	○	○	○	○

Mains lead supplied   Switch Live

Remove link plug from programmer terminal strip.

Floor mounted

Suitable for use on sealed systems using optional kit
Heat exchanger material, mild steel

Worcester Heatslave combination 012–14, 015–19, 020–25

Mains lead supplied. External controls, such as a time clock with voltage free contacts or room thermostat should be connected across terminals 2–4 after removing link. Room thermostat neutral can be connected to terminal N2.

Floor mounted

Not suitable for use on sealed systems
Heat exchanger material, mild steel

Worcester Heatslave 2+ combination G40, G50

Mains lead supplied. External controls, such as a time clock with voltage free contacts or room thermostat should be connected across terminals 2–4 after removing link. Room thermostat neutral can be connected to terminal N2.

Floor mounted

Not suitable for use on sealed systems
Heat exchanger material, mild steel

Installing and Servicing Domestic Heating Wiring Systems and Control

Control wiring connections for when connected directly into burner control box

Control box	Perm live	Control live	N	E	Link information
Danesmoore TSV	L	HW ON-4 CH ON-6	N	E	Remove Links A–1 and A–2
Danfoss BHO 1		6*	N	E	
Danfoss BHO 15		9+	7	E	
Danfoss 57 F		1	3	E	Link 2–3
Danfoss 57 H		6	1	E	
Danfoss 57L		5	3	E	
DS 220		2	N	E	Link N–L1
Elestra		2	1	E	
Honeywell Protectorelay		1	2	E	
Landis & Gyr LAB 1		1+	2	E	
Landis & Gyr LOA 21		1+	2	E	
Nu-way ZLO		6	1	E	
Nu-way ZL2D		1	4	E	
Petercem MA28		10	4	E	
Satchwell DG		1	2	E	
Satronic TF 701 B		9+	8	E	
Satronic TF 830 N		9+	8	E	
Selectos D42		6	1	E	
Selectos JSS 1		6+	1	E	
Selectos JSS 2#		8	2	E	
Stewart TSV		as Danesmoor TSV			
Teddington DAS	L	11	N	E	Remove Link L–11
Thermoflex MC50		4	2	E	
Trianco TSV		as Danesmoor TSV			

* via limit thermostat
+ via control and limit thermostats
\# connections into motor burner starter
Note: the earth connection may be a nut and bolt stud or screw into casing

10

Ancillary controls

Contents

ACL Clockwatcher

Danfoss BEM 4000 Boiler Energy Manager
Danfoss BEM 5000 Boiler Energy Manager

Dataterm Optimiser

Drayton Theta-Autotherm Compensator System

Gledhill Boilermate
Gledhill Cormorant
Gledhill Gulfstream

Honeywell AQ6000 Compensator System
Honeywell Y604A Sundial Plans
Honeywell Y605B Sundial Plans

Pegler Belmont tec Multiple Zone Programmer

Randall EBM 2.1 Boiler Efficiency Control

Ringdale 702 Boiler controller
Ringdale 802 System Controller

Sunvic Clockbox and Clockbox 2

Installing and Servicing Domestic Heating Wiring Systems and Control

ACL Clockwatcher

The ACL Clockwatcher is a thermistor-controlled timing device which intercepts the 'heating-on' signal from a clock or programmer and then delays switch-on until the correct time under the prevailing temperature conditions. It is a single sensor optimizer, monitoring either inside or outside temperatures – not both. Three different sensors were available.

The ACL Clockwatcher is wired in and the existing clock or programmer set to provide warmth when it is required under the coldest expected conditions. The time period between switch-on and occupancy (preheat period) is then programmed into the Clockwatcher by means of the slotted setting knob on its front panel, calibrated from 1 to 6.

Adjustment of this knob changes the base time delay range over which the Clockwatcher operates. At setting 1 the delay range is from 10 minutes to 1 hour over the full operating temperature range, while at setting 6 the delay is from 1 hour to 6 hours. This adjustment, and the choice of sensor types, gives the user the ability to match the control to his particular requirements. A manual override is also provided to enable the heating system to be turned on immediately should circumstances demand.

The type of sensor chosen will determine where the Clockwatcher is located. If the outside or remote sensors are chosen, then the Clockwatcher can be fitted virtually anywhere. If the internal sensor is selected, then the Clockwatcher should be mounted in a position normally suitable for a room thermostat. The sensors should be connected into the Clockwatcher as shown in Figure 10.2. A wiring diagram for gravity hot water, pumped central heating is also shown in Figure 10.1.

Detector connections

	Violet lead	Orange lead	Sensor leads
Outside detector	7	10	8 and 9
Detector in unheated room	10	9	8 and 10
Detector in heated room	7	9	8 and 10

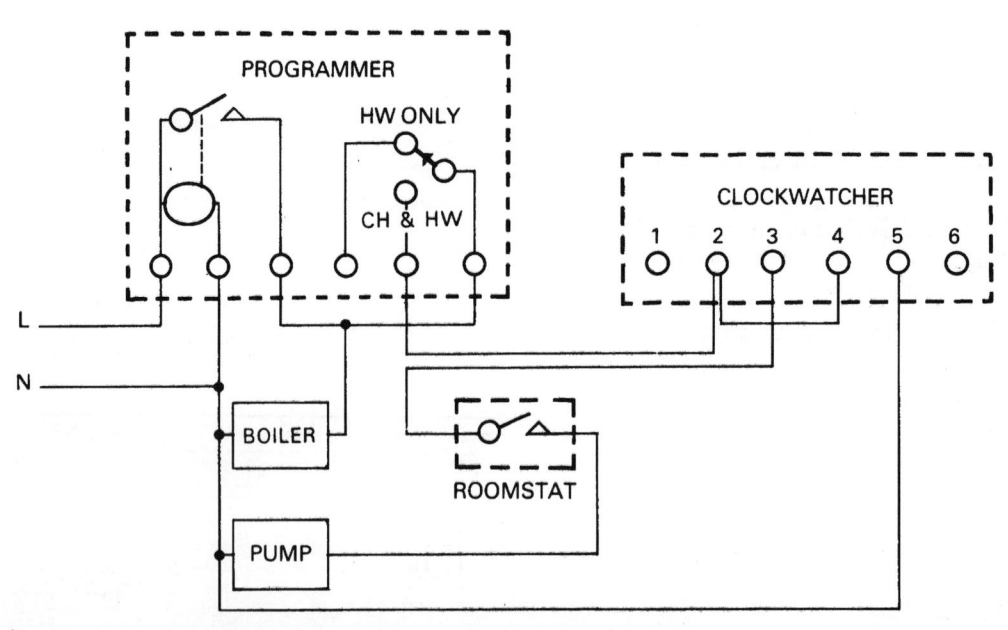

Figure 10.1

Ancillary controls

Setting the Clockwatcher

As an example, if the programmer is set to switch the heating on 2 hours before the house is to be warm then, using a screwdriver, adjust the Clockwatcher to position 2. Where the programmer is set for 1½ hours before, adjust Clockwatcher between 1 and 2, and so on. Where the heating is required immediately, press the override button on the Clockwatcher, ensuring the programmer is in an 'on' position. The Clockwatcher will automatically reset as the programmer switches on/off.

Testing the Clockwatcher

With the power on, set programmer for hot water and heating on constant, and check that the boiler comes on for hot water. The heating should not respond until the time delay has expired. Check the heating using the override facility.

Figure 10.2 *Gravity hot water, pumped central heating*

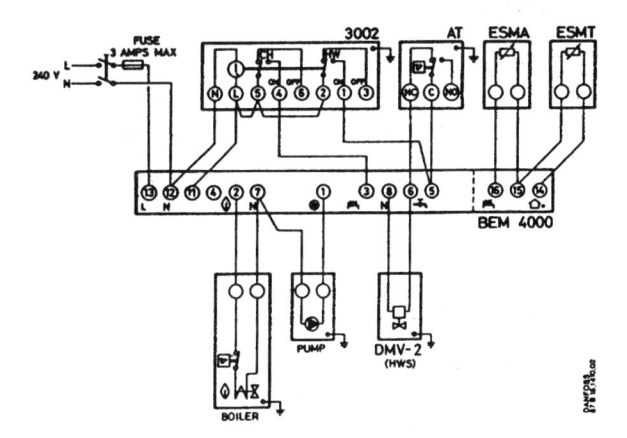

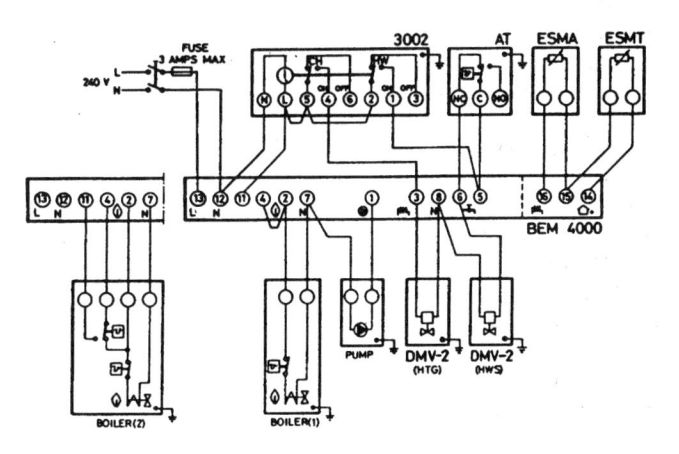

Figure 10.3 *Fully pumped system with two motorized zone valves*

Danfoss BEM 4000 Boiler Energy Manager

The BEM is an electronic controller which can be added to almost any central heating system to eliminate unnecessary boiler cycling and improve boiler seasonal efficiency.

The BEM 4000 consists of an electronic controller, an outdoor temperature sensor and a strap-on flow temperature sensor. It can be used in conjunction with a room thermostat (utilized as a frost stat) and TRVs. Boiler performance is improved when the flow temperature is varied according to the outside temperature. The outside temperature sensor should be mounted on the coldest elevation of the building away from opening windows and the boiler flue. The flow sensor should be fitted within 6 inches of the boiler outlet on the flow pipe, or on low water content boilers it can additionally be fitted between the boiler and the by-pass. A cylinder thermostat should be fitted as normal. All

165

Figure 10.4 *Fully pumped system with mid-position valve*

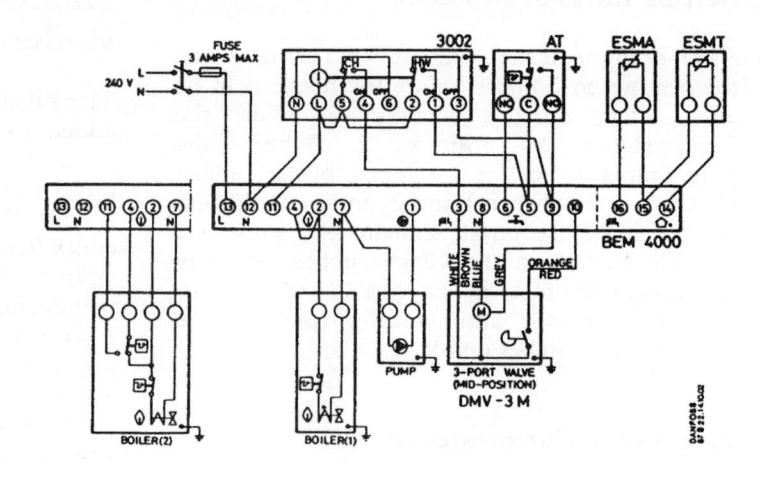

diagrams show Danfoss programmer and thermostats, but others can be used. In all cases, if a room thermostat is to be utilized as a frost or low-limit thermostat, connect to Terminals 3 and 11.

In Figures 10.3 and 10.4 two boilers are shown – one with pump over-run and one without. Select diagram as appropriate, remembering to insert Link 2–4 when using boiler without pump overrun.

Commissioning the BEM 4000
Hot water service

Set the programmer to hot water constant, central heating off, and cylinder thermostat to maximum. Set boiler to the design flow temperature of 82°C.

Lamps 1 and 2 will illuminate and boiler and pump will start, except in gravity hot water systems, in which the pump will not start. The hot water valve will open.

Central heating service

Set programmer to hot water off, central heating constant and boiler as above. Where a heating motorized valve is fitted, it will open. Illumination of lamp 3 signifies that the boiler has stopped and after a time lamp 1 will go off, indicating pump stopped.

After testing, set the programmer and cylinder thermostat to the required settings.

Fault finding

When the 'on' lamp on the front of the BEM 4000 does not illuminate, check the power supply. There should be 240V across Terminals 12 and 13. If this is OK, check the built-in fuse on the rear of the BEM 4000, and if blown, there is a spare attached to the rear of the unit. If this fuse is OK the fault is probably within the electronics, and the BEM 4000 plug-in front plate should be replaced.

Where a zone valve will not open, check that 240V exists at the valve leads. If OK change valve, if not check programmer and other wiring.

In the event of difficulties, the BEM 4000 can be overridden by inserting a link between Terminals 14 and 15, ensuring that the power is turned off first.

Danfoss Randall BEM 5000 Boiler Energy Manager

The BEM 5000 is an electronic control, which, when added to almost any central heating system, reduces unnecessary boiler cycling and improves boiler seasonal efficiency. It is compatible with most popular control systems including radiator thermostats, programmers, motorized valves and even room thermostats, although in systems having radiator thermostats it is recommended

Ancillary controls

Figure 10.5

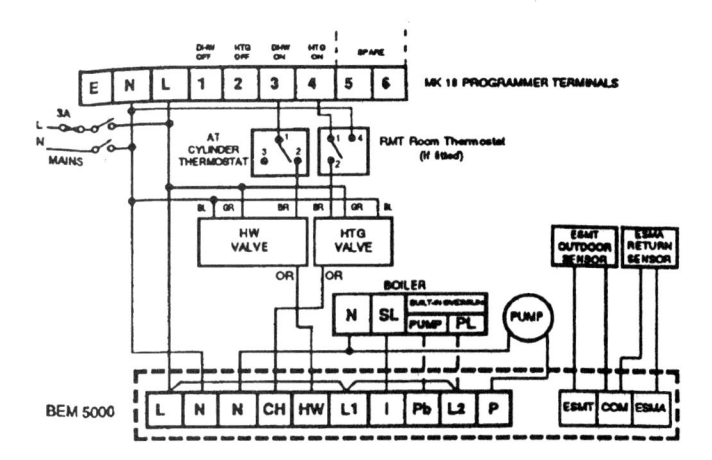

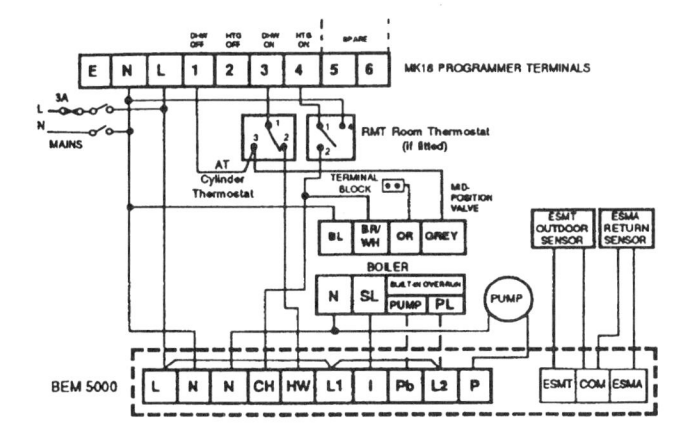

Figure 10.6

that the room thermostat be turned up to maximum or re-wired as frost thermostats. Radiator thermostats are recommended, but are not mandatory. If they are fitted, then a by-pass valve should be fitted. If it is impractical to fit a by-pass valve, then one radiator should be left uncontrolled.

The BEM 5000 measures outdoor temperature and varies the temperature of water flowing to the radiators accordingly. This significantly improves boiler performance. A water temperature sensor is used to monitor temperature of water returning to the boiler. Any change in load on the system is measured and the boiler is controlled accordingly.

To minimize energy loss from the boiler case and flue, the pump runs on after the boiler has stopped, circulating all useful heat to the system. When no more heat can be extracted, the pump is stopped. The BEM 5000 integrates the operation of heating and hot water. During periods of hot water demand, water temperature flowing to both water and heating circuits is boosted to boiler thermostat settings to ensure rapid recovery. Figures 10.5 and 10.6 show the BEM 5000 wired into a fully pumped system, one with two 2 port motorized valves and one with a 3-port, mid-position valve. It can also be utilized on a combination boiler.

167

Installing and Servicing Domestic Heating Wiring Systems and Control

Dataterm optimiser

Dataterm is a microprocessor-based energy management system and can be used as part of a new system or to replace traditional room thermostats and programmer. The fully automatic optimum start function takes into consideration the current weather and calculates the boiler start time eliminating fuel waste through fixed boiler start times. Dataterm is suitable for any conventional oil or gas wet central heating system complete with hot water control, as well as other applications.

The control pack consists of two units, a low voltage programmable room thermostat and a controlling power pack located in a convenient position, e.g. adjacent to the boiler or motorized valves.

```
              O    + VE
DATATERM
              O    OV

              O
                   12V AC
              O

              O
                   12V DC
              O
```

Drayton Theta Autotherm

The Drayton Autotherm was one of the first compensator systems specifically designed for the domestic property. It consisted of a controller which was similar to a programmer, a mixing valve and two temperature sensors – one to detect outside temperature and one to detect water temperature to radiators. Additional thermostats could be added to provide control of hot water, frost protection, etc.

As this unit is no longer available brief wiring instructions are given should the need for fault finding arise, together with a schematic diagram of how the system works (Figures 10.7, 10.8).

Note: NO = normally open
       NC = normally closed
       N = neutral

1. Gravity hot water, pumped central heating

O	O	O	O	O	O	O	O	O	O	O	O	O	O	O	O	O	O
EARTH	DEM	COM	N	L		NO	N	E	NC	NO	N	E	NC	NO	N	E	NC
	CYL STAT		MAINS			BRN	BLU	GY		L	N	E		L	N	E	
						2 PORT HOT WATER VALVE				PUMP				BOILER			

2. Fully pumped 3-port mid position motorized valve

O	O	O	O	O	O	O	O	O	O	O	O	O	O	O	O	O	O
EARTH	DEM	COM	N	L		NO	N	E	NC	NO	N	E	NC	NO	N	E	NC
	CYL STAT		MAINS			ORA	BLU	GY	GRY	WHI				L	N	E	
						3 PORT MID POSITION VALVE								BOILER AND PUMP			

3. 2 × 2 port motorized valves

O	O	O	O	O	O	O	O	O	O	O	O	O	O	O	O	O	O
EARTH	DEM	COM	N	L		NO	N	E	NC	NO	N	E	NC	NO	N	E	NC
	CYL STAT		MAINS			BRN	BLU	GY		BRN	BLU	GY		N	E		
						HOT WATER VALVE				HEATING VALVE				BOILER AND PUMP			

Note: Connect boiler and pump lives with motorized valve oranges in separate connector. Connect morotized valve greys into mains L.

Ancillary controls

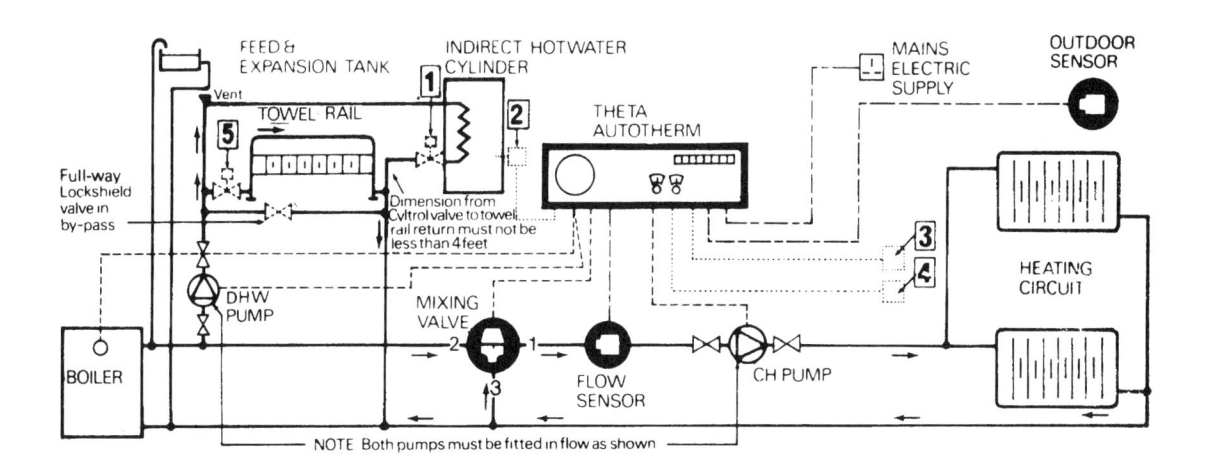

Figure 10.7

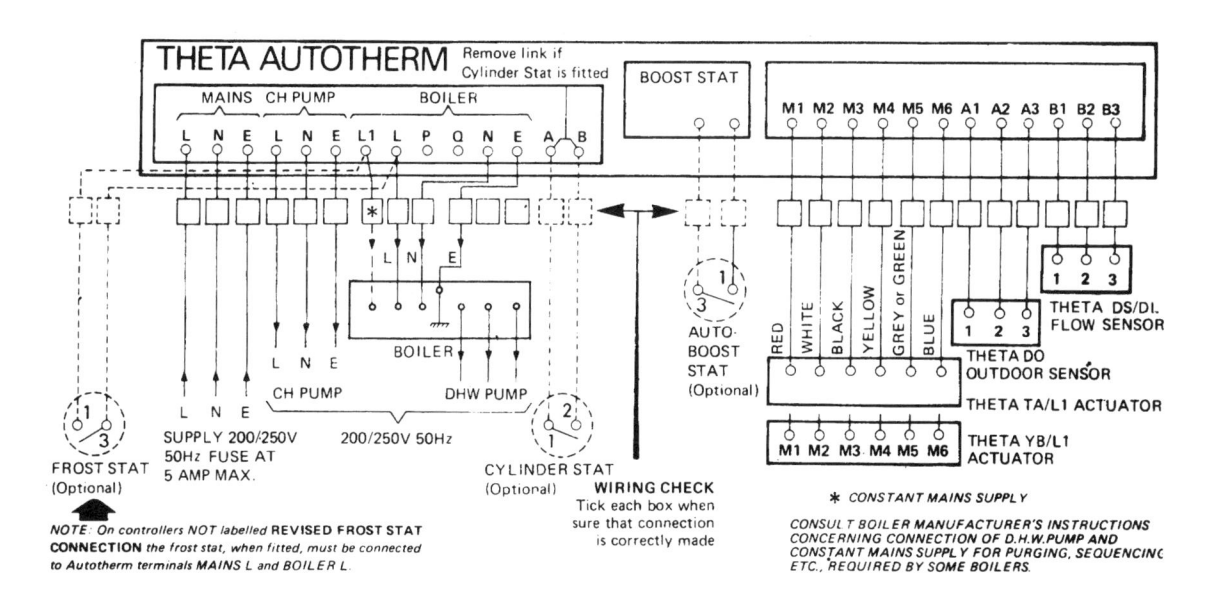

Figure 10.8 *Drayton Autotherm on a gravity hot water, pumped central heating system*

169

Gledhill Boilermate

The Boilermate is used to provide improved hot water and space heating, as well as an unvented mains pressure hot water supply systems, for use with any boiler sited remotely. A boiler of any output **up to a maximum of 80 000 BTU** can be linked to any model of Boilermate, and the deciding factor is the hot water requirement. The boiler thermostat must always be set at maximum.

The principle of Boilermate is to separate the heat generator (e.g. the boiler) from heat emitters by a thermal store, which evens out the fluctuating demands for heating and hot water. Thus by storing energy produced when demand is low and discharging it when demand is high (i.e. during warm-up or when hot water is drawn off), a smaller boiler can be used. An important feature of this design is that hot water can be supplied directly from the mains at conventional flow rates without the need for temperature and pressure safety relief valves or expansion vessels, at mains pressures of between 1 and 5 bar.

The thermal store contains primary water which is maintained at a temperature between 80°C and 85°C at the top of the sore. It is efficiently insulated with rockwool insulation in a galvanized steel case to minimize standing losses.

The radiator circuit is controlled by the heating circuit pump through the action of room thermostat and time clock.

For fuel economy and best boiler performance, the system should be designed so that gravity circulation does not take place in the heating system when the pump is not running. It is often necessary to fit a check valve to be certain of preventing this.

The cylinder thermostat is the principal control mechanism and the correct type is therefore important. This will be supplied and pre-set as part of the control package.

A delay timer is pre-wired into the Boilermate wiring loom. **Link 1 must be removed for boilers which have pump over-run.**

The delay timer is activated on boilers without pump over-run by the link between 4 and 8. This aids system economy by removing all the hot water from the boiler each time it shuts down.

Gledhill Cormorant

Cormorant is a gas-fired appliance for supplying both wet central heating and hot water for dwellings with a design heat loss of up to 4kW open flue and 5kW balanced flue. The principle is to separate the heat generator (e.g. the boiler) from heat emitters by a thermal store, which evens out the fluctuating demands for heating and hot water. Thus by storing energy produced when demand is low and discharging it when demand is high (i.e. during warm-up or when hot water is drawn off) a much smaller appliance can be specified. An important feature of this design is that hot water can be supplied directly from the mains at conventional flow rates without the need for temperature and pressure safety relief valves, or expansion vessels at mains pressures of between 1 and 5 bar.

Cormorant is a thermal storage system which comes complete with its own boiler in open-flued form and in which circulation between heat source and thermal store is by gravity circulation.

Cormorant models are designed to be fed directly from the mains and their performance is directly related to the adequacy of the cold supply to the dwelling. This must be capable of providing for those services which could be required simultaneously, and the maximum demand should be calculated. Cormorant will operate at pressures as low as 1 bar, which must be available when local demand is at its maximum, but the preferred range is between 2 and 5 bar. As a general guideline, a 15mm copper supply is sufficient for a 1 or 2 person dwelling and therefore usually adequate for a Cormorant installation.

Maxol 12D balanced flue circulator, or the Baxi WM20/3RS balanced flue boiler, or the Maxol Superstream open flue circulator, are recommended. However, Cormorant thermal stores can be used with back boiler unit, such as the Maxol 15 BBU. The Cormorant is designed specifically for use with gravity circulation between the heat source and the thermal store.

The boiler can be controlled by a simple time switch or on/off control to supply the thermal store. The heating pump can be controlled by a time clock, room thermostat, or both, or again a simple on/off switch, as it only takes minutes for the radiators to get warm because the thermal store is always hot.

Gledhill Gulfstream

The Gulfstream is a gas-fired, open flue appliance for supplying wet central heating and means pressure hot water for dwellings with a design heat loss of up to 5kW. The principle is to separate the heat generator (i.e. the boiler) from the heat emiters by a thermal store of approx. 120 litres capacity. This evens out the fluctuating demands for heating and hot water.

An important feature of this design is that hot water can be supplied directly from the mains without the need for safety controls (e.g. pressure and temperature safety relief valves, etc.). This is achieved by passing the mains water through high performance twin heat exchangers complete with built-in, self-recharging expansion chamber.

Another important feature of this system is that, because the thermal store acts as a buffer between the boiler and the heating system, any variety of space heating control systems can be used.

Gulfstream is not suitable for sealed systems. The appliance has a built-in F&E tank and therefore all system components (e.g. radiator) must be at least 250mm (10") below the water level of the tank.

If an inhibitor is to be used, it must be of a British Gas approved type and suitable for copper tube boilers. The system should be designed so that gravity circulation does not take place in the radiator circuit when the heating pump is off.

Gulfstream is fitted with a low limit store thermostat. This gives priority to hot water, therefore after a heavy draw-off (e.g. bath), the central heating will not be served for a short period whilst the appliance is reheating the thermal store.

A feature of the Gulfstream system is that because water is stored at a high temperature, the radiators become warm within minutes of switching on the pump. This rapid response means that the time clock can be set to switch on the heating approximately 10 minutes prior to requiring heat.

Honeywell AQ 6000

The AQ 6000 is an outside temperature compensator system for use in domestic properties and can be used on new installations or to upgrade existing systems, making use of existing valves that may be installed. The AQ 6000 is available in three packs – an Upgrade pack, a Standard pack or a Modulating pack. The Modulating pack is for use on boilers of 90 000–150 000 BTU and provides control of the heating circuit by modulating a mixing valve. There are five systems – A, B, C, D and E. Systems D and E use the modulating mixing valve and are not shown.

The AQ 6000 consists of a control unit, a room unit and outside, water supply and domestic hot water sensors. All wiring is terminated in the control unit along with boiler, pump and any valve wiring.

A built-in start-up operating sequence allows the system wiring to be tested. Before turning on power to the system, remove the room unit from its mounting bracket. Turn on the power and the control unit will power the valve, pump and boiler.

When the system is operational actual temperatures can be displayed on the room unit. When the green enquiry button is pressed, 'T1' will be displayed at the left side of the display and the temperature in °C on the right. Repeated use of this button will display other system temperatures. The codes are listed in Table 10.1. To restore the display to its normal operation press the enquiry button repeatedly until the 'T' is no longer displayed and in its place is the current weekday number.

Table 10.1

Code	Temperature measured
T1	Room temperature
T2	Boiler/mixed water temperature
T3	Outside temperature
T4	System C – not used (will display zero °C)
	Systems A and B – domestic hot water temperature

When the room unit is connected and the supply switched on, an indication of any faults within the system can be displayed by pressing any four buttons on the unit. A fault code is displayed as an 'F' followed by a number. The different numbers correspond to the faults listed in Table 10.2. No 'F' on the display means that no fault has been observed by the controller. To clear the fault code, press any button.

Installing and Servicing Domestic Heating Wiring Systems and Control

Table 10.2

Code	Indicates	Possible causes	Corrective action
F1	How water has not reached 30°C within 40 minutes of start-up	Boiler not firing	Check wiring
			Check appliance function
		Pump not running	Check wiring
		Valve being driven closed	Reverse leads 1 and 2 on the actuator *or* on the boiler unit
		Valve in closed position	Check wiring
			Check actuator
F2	Boiler/mixed water temperature sensor fault	Faulty sensor wiring	Check wiring for open or short circuits
		Faulty sensor	Change sensor
F3	Outside temperature sensor fault	Faulty sensor wiring	Check wiring for open or short circuits
		Faulty sensor	Change sensor
F4	Failure of the communications link between the room and the boiler room	Room unit not securely seated in the mounting bracket	Check seating of room unit in its bracket
		Faulty wiring	Check for open or short circuits
		Wire connected the wrong way round	Reverse wiring connection at the boiler unit or room unit
		No power at boiler unit	Check boiler supply and 50mA fuse on boiler unit board
F5	Limit or domestic hot water temperature sensor fault	Faulty wiring	Check wiring for open or short circuits
			On systems C and D, check that the resistor is still in place between terminals 3 and 4 of the boiler unit
			On other systems, check that the resistor has been removed
		Faulty sensor	Change sensor

Ancillary controls

System A: Boiler control with domestic hot water control using two motorized zone valves

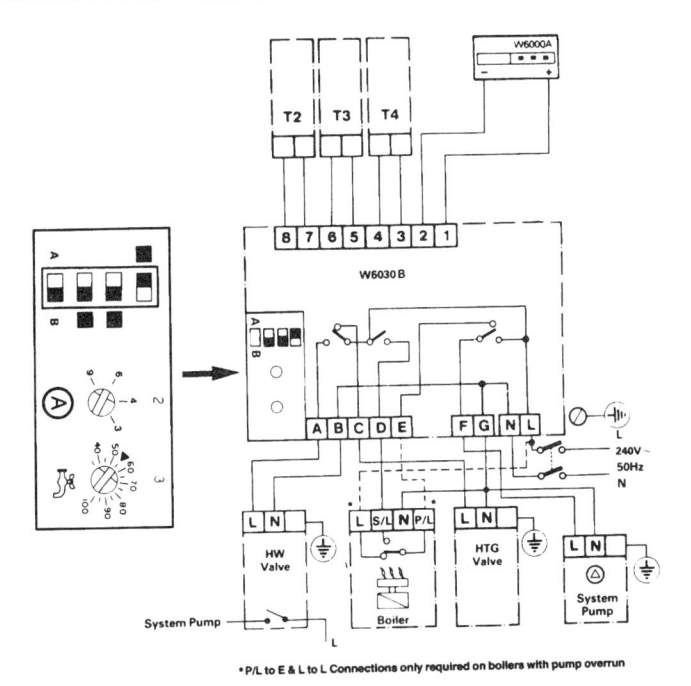

The following adjustments must be made **before** power is applied to the system:

1. System selection switches.
2. Burner cycles per hour (3 to 9), typical setting: 6.
3. Domestic hot water (35 up to 100°C), typical setting: 55°C.

Note: The switch A–B should be in position A for the domestic hot water to be off when the system is controlling at the Economy level. The switch should be in position B for the domestic hot water to be continuously serviced.

P/L to E & L to L Connections only required on boilers with pump overrun

Figure 10.9

System B: Boiler control with domestic hot water control using a 3-port motorized valve

The following adjustments must be made **before** power is applied to the system:

1. System selection switches.
2. Burner cycles per hour (3 to 9), typical setting: 6.3.
3. Domestic hot water (35 to 100°C), typical setting: 55°C).

Note: The switch A–B should be in position A for the domestic hot water to be off when the system is controlling at the Economy level. The switch should be in position B for the domestic hot water to be continuously serviced. Wiring shows a priority valve. Where a mid-position valve is used the White (or Brown) and Grey go to terminal C.

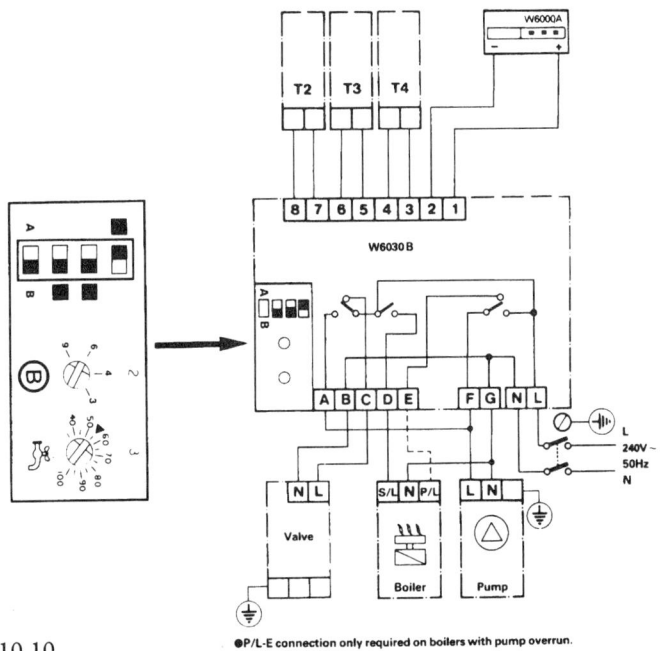

●P/L-E connection only required on boilers with pump overrun.

Figure 10.10

173

Installing and Servicing Domestic Heating Wiring Systems and Control

System C: Boiler control without domestic hot water control

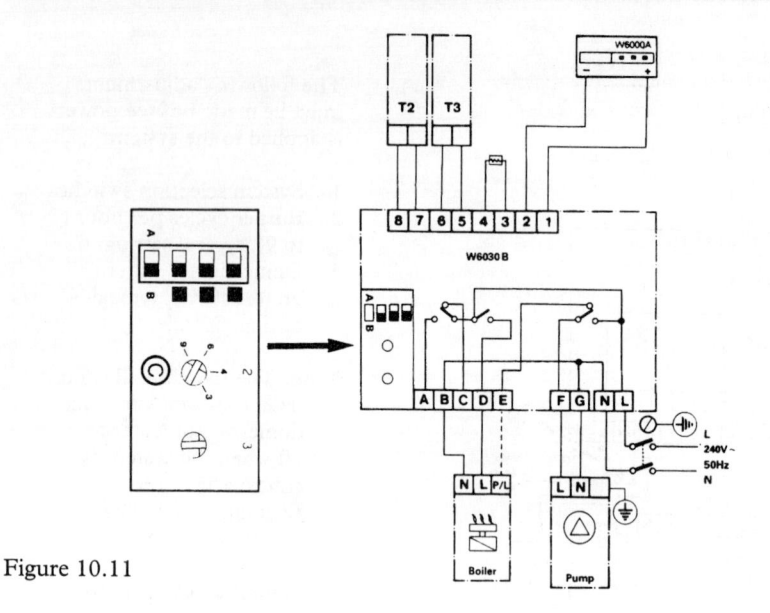

The following adjustments must be made **before** power is applied to the system:

1. System selection switches.
2. Burner cycles per hour (3 to 9), typical setting: 6.

Note: Only recommended for systems up to 70 000 btu. Domestic hot water sensor, T4, is not used.

Figure 10.11

Honeywell Y604A Panel and Timed Sundial Plans

The Y604A Panel and Panel Timed Sundial Plans are fully assembled, pre-plumbed and pre-wired control sets and comprise of either a Honeywell S or Y Plan, or Timed S or Y Plan, plus pump and isolating valves in a multipoise carrier bracket.

Table 10.3   *External wiring connections – panel plans*

		S Plan terminal	Y Plan terminal
Room thermostat	Common Demand Neutral	4 5 N	4 5 N
Basic boiler	Live Neutral Earth	7 N E	8 N E
Boiler with pump overrun	Remove link 7–9 in junction box. Live Neutral Earth Boiler on Pump live	L N E 7 9	L N E 8 9
Programmer	Live Neutral Heating on Hot water on Hot water off	L N 4 6 –	L N 4 6 7
Frost thermostat	Common Demand	1 5	1 8

Ancillary controls

Table 10.4 *External wiring connections – panel timed plans*

		S Plan terminals	Y Plan terminals
Room thermostat	Common	4	4
	Demand	5	5
	Neutral	N	N
Basic boiler	Live	7	8
	Neutral	N	N
	Earth	E	R
Boiler with pump overrun	Remove link 7–9 in junction box		
	Live	L	L
	Neutral	N	N
	Earth	E	E
	Boiler on	7	8
	Pump live	9	9
Frost thermostat	Common	1	1
	Demand	5	8

Honeywell Y605B Panel Link-Fuel Timed Sundial Plan

The Y605B Panel Link-Fuel Timed Sundial Plan is an assembled, pre-plumbed and pre-wired control set. Designed for use with fully pumped wet central heating systems linking together a solid fuel boiler with either a gas, oil or electric boiler. Full time and temperature control of the system is achieved by use of the programmer, room and cylinder thermostats.

The system selector box offers the choice of heating by means of solid fuel only, where the link-fuel boiler will not be used, or as a link-up system where both the solid-fuel and link-fuel boilers will be used.

A second selection allows choice between dissipating excess heat either into the heating or the domestic hot water circuits. It is recommended that heating be selected during the winter and hot water in spring or autumn, whenever the solid fuel appliance is in use.

When commissioning, set the required on/off times of the system on the programmer, set required temperature for domestic hot water, and the room thermostat to the required room temperature. The triple aquastat limit thermostat is factory set but may need to be adjusted.

Control operation

A. When solid fuel only has been selected. The link-fuel valve will be closed and the solid fuel valve will be open and the system will operate as follows:

1. No demand within the system slumber valve open to allow full bore gravity circulation to the slumber circuit. Safe operation of solid-fuel appliance without overheating of space or domestic hot water. Heating valve and domestic hot water valve closed, pump off.
2. Heating only: pump on, slumber and hot water valves closed, heating valve open to allow heating to be provided. Water pumped around the slumber circuit through adjustable gate valve.
3. Hot water only: pump on, slumber and heating valves closed, hot water valve open to allow circulation to the domestic hot water. Water pumped around the slumber circuit through adjustable gate valve.
4. Both heating and hot water: pump on, slumber valve closed, heating and hot water valves open to allow circulation throughout the whole of the system. Water pumped around the slumber circuit through the adjustable gate valve.

5. If the temperature of the water exceeds the high limit setting at any time the safety controls will take effect. Then pump on, slumber valve closed whilst either heating or hot water valve will open to allow for excess heat to be dissipated into either the hot water or heating and slumber circuits, as selected. The high limit thermostat will automatically reset when the temperature of the water has reduced and the system will return to normal operation.

B. **When the link-fuel has been selected the system operates in conjunction with the limit thermostats:**

1. When flow temperature from solid fuel appliance is below the mid limit, the link-fuel boiler is brought into use. The slumber valve is closed, link-fuel and solid-fuel valves open and either heating, or hot water, or both, valves open to satisfy demand of the system. Water is supplied to the slumber circuit through the adjustable gate valve and both the pump and link-fuel boiler are on. Operation of the system is switched between the solid fuel and link fuel by the mid limit thermostat.
2. When flow temperature from solid fuel appliance is between the mid and high limit temperatures then the link-fuel boiler is switched off. Full use is made of the output from the solid fuel appliance only, hence achieving maximum use of the solid fuel appliance and economizing on other fuels. Link-fuel valve closed.
3. If the flow temperature from the solid fuel appliance exceeds the high limit setting then the link-fuel boiler is switched off. Slumber and link-fuel valves close, solid fuel valve opens as well as either heating or hot water valve, depending upon selection of either heating or hot water for dissipation for excess heat. Pump is on to remove excess heat through the chosen circuit and also through the slumber circuit via the adjustable gate valve.
4. The low limit thermostat provides control of the solid fuel appliance at low temperature.
   (a) Temperatures below the low limit setting, solid-fuel valve will be closed.
   (b) Temperatures above the low limit setting, solid-fuel valve will be open.
   (c) Any requirement for heating or hot water will be met by the link-fuel boiler exactly as Item 1 above.

Note: When pump is off slumber valve is open
When pump is on, slumber valve is closed

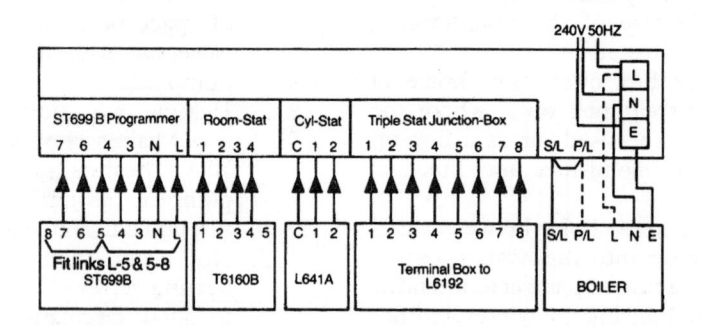

Figure 10.12

When using pump overrun boiler do not fit link S/L to P/L but connect L to L and P/L to P/L.

Ancillary controls

Pegler Belmont tec Multiple Zone Programmer for TRVs

The Belmont tec is a seven-day programmer designed to directly control thermostatic radiator valves (TRVs) by means of a bell wire connection. When the heat is required in an area of a building the TRVs will functions as a standard TRV. When heat is not required the tec controller will send a current along the bell wire to heat up the TRV head and shut off the valve.

The system has been designed to complement any wet central heating system without plumbing modifications to existing pipework when retro-fitting and required less pipework when designing new heating installations.

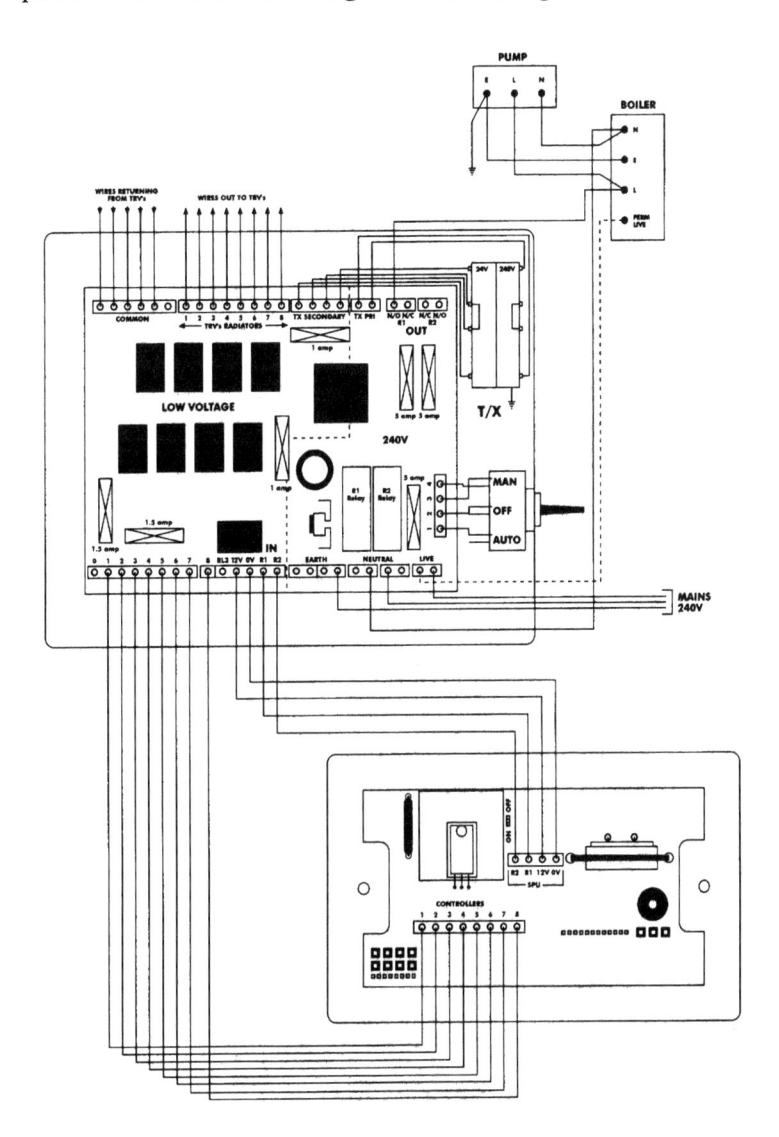

Figure 10.13

The system is ideally suited for small hotels/guest houses, retirement homes, nursing homes, large domestic houses and larger premises.

Randall EBM 2.1 Boiler Efficiency Control

The EBM 2.1 Boiler Efficiency Control is designed, with one simple user adjustment, to maximize the efficiency of gas fired wet domestic central heating systems. This is for systems which incorporate time and temperature controls (TRV or room thermostat and cylinder thermostat) and motorized valves. Suitable for both new and existing installations incorporating conventional or condensing boilers in either gravity or fully pumped systems.

The following pages show various systems and their wiring diagrams, however, before turning to these it is important to note the following:

1. The EBM 2.1 should be mounted where it is clearly accessible.
2. The OTS outside sensor should be mounted on an outside wall on the coldest side of the house (usually north facing) about 18 inches below eaves height and away from chimneys, flues or windows, etc. Connection must be made with cable suitable for 240V as follows:

   Terminal 1 on the OTS to terminal 11 on the EBM wallplate
   Terminal 2 on the OTS to terminal 10 on the EBM wallplate

3. The BRS boiler return sensor should be mounted on the common return to the boiler between the boiler and by-pass. In gravity hot water systems it should be mounted on the heating return. Connect to EBM as follows:

   Brown lead to terminal 12 on the EBM wallplate
   Blue lead to terminal 10 on the EBM wallplate

4. The cylinder thermostat and any motorized valves shown are not included in the kit.
5. The pump should be connected to terminal 5 of the EBM in all cases, even if the boiler has its own pump connection or pump overrun facility.
6. In gravity hot water systems, it is necessary to cut an internal link in the EBM by prizing off the back cover and cutting the grey link wire located inside the housing on the left.
7. The boiler thermostat should be set to maximum.

Commissioning the EBM 2.1

1. Set the response rate setting on the EBM 2.1 to 10, **except for condensing boilers**, where the response rate should be set at 5.
2. Set the boiler thermostat to maximum.
3. Set the cylinder thermostat to maximum.
4. Set the room thermostat (if fitted) to maximum.
5. Turn on the power to the system. The 'Saving' light will illuminate.
6. Set the programmer on HW 'ON'. The 'HW' and 'Pump' lights will illuminate. The boiler will fire and the pump will run. The 'Saving' light will switch off.
7. Allow the system to warm for about 15 minutes and turn the cylinder thermostat to minimum. The 'HW' light will switch off and the boiler will stop. The pump (and 'Pump' light) may continue to run depending on the return temperature (see note above for gravity primary systems). The 'Saving' light will switch on.
8. Set the programmer to CH 'ON'. The 'CH' light will illuminate. The 'Saving' light will switch off and the boiler and pump will run.

   Note: If the outside temperature is greater than approximately 17°C, it will not be possible to run the CH service as outlined above as it will not switch on. If this is the case, disconnect the sensor lead from terminal 11 on the wallplate which will force the CH service on. After checking, the sensor lead should be reconnected to terminal 11.

9. When the system is up to temperature, the boiler will stop and the 'Saving' light will illuminate. The pump will continue to run. When the system temperature drops sufficiently the boiler will fire again and the 'Saving' light will switch off.

Ancillary controls

Figure 10.14 *Fully pumped system with two motor open, motor closed motorized valves*

Figure 10.15 *Fully pumped system with two spring-return motorized valves*

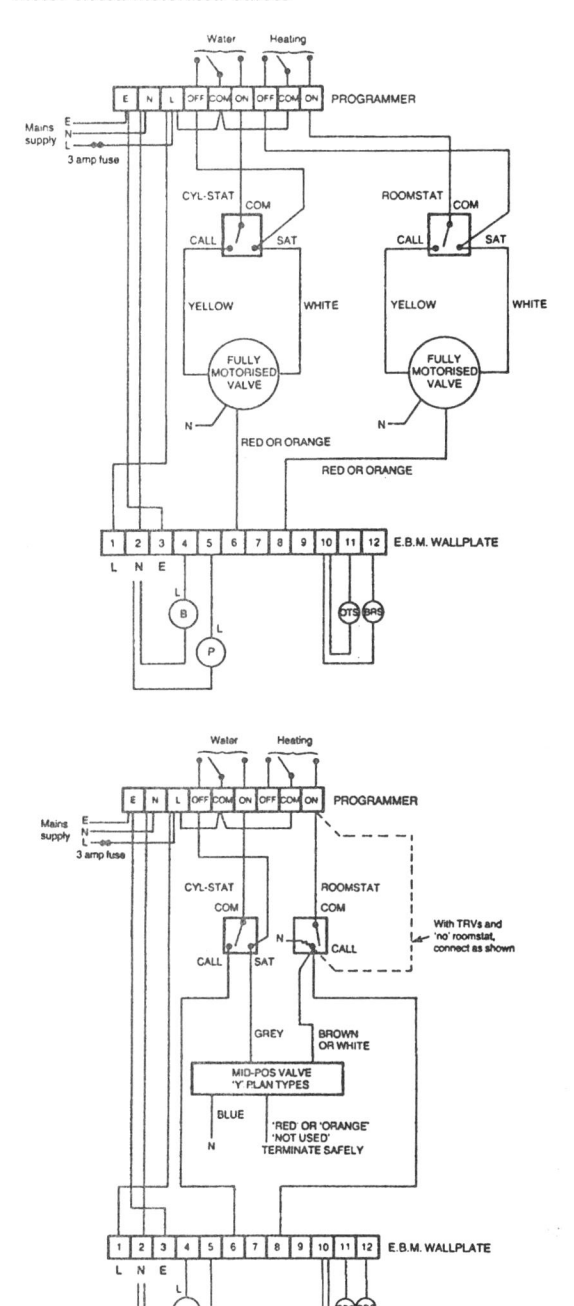

Figure 10.16 *Fully pumped system with mid-position motorized valve*

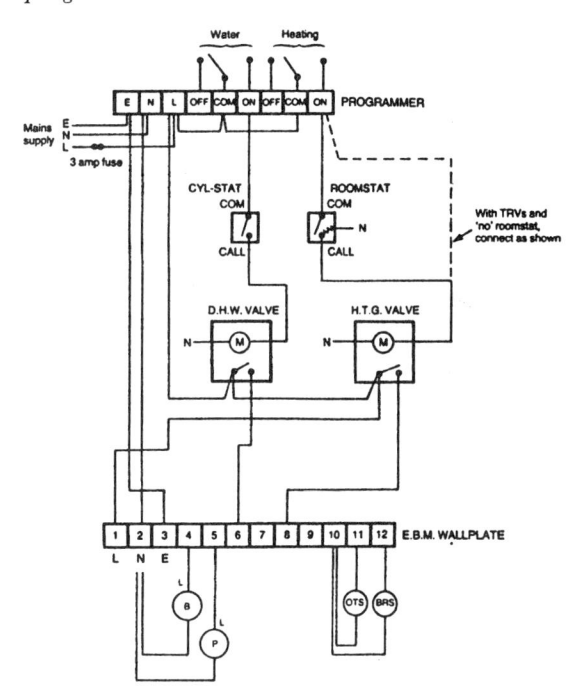

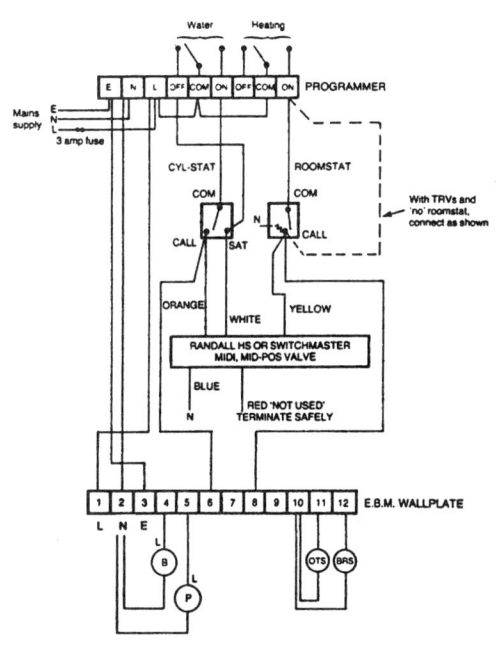

Figure 10.17 *Fully pumped system with Switchmaster 'Midi' or Randall HS mid position valve*

179

Ringdale 702

The 702 Boiler Controller is intended for both existing and new central heating systems. When fitted on systems which do not have a hot water cylinder thermostat, the 702 will achieve a substantial fuel saving by preventing spurious boiler firing. In most types of systems the 702 will provide a high degree of user convenience through easily accessible temperature setpoints and improved accuracy and stability.

The 702 can be used in most domestic central heating systems and can directly switch pumps, a boiler and two or three port valves in various combinations. It senses temperatures in the house and on two points on the hot water cylinder, and energizes the boiler only when either the hot water cylinder or the house require heat. A switch on the 702 selects which of the two sensors on the hot water cylinder is active and this permits the user to heat either a full cylinder or just the top half. This simple feature makes a significant contribution to the saving which can be achieved. An eternal electronic or mechanical programmer is required for presettable time-slots. The sensor leads can be extended up to 30 metres with standard 2-core mains cable, without affecting accuracy.

Ringdale 802

The 802 is a sophisticated microprocessor-based heating system controller. It can control up to three zones, any of which may be hot water (HW) or heating (HTG). The 802 may be programmed to maintain a profile of different temperatures at different times for each day of the week, with only one restriction: a maximum of 50 time/temperature points per week. Each zone may be programmed independently of the others. This gives unprecedented flexibility which is particularly beneficial in larger installations where the heating can be programmed to exactly match varying requirements.

A Manual mode provides constant temperature control on all three zones, for a specified time period (1–24 hours), after which the 802 reverts to the automatic programme.

A Holiday mode maintains a minimum temperature (adjustable 0–20°C) in all three zones, until a specified date is reached. The date is set in the common DAY/MONTH/YEAR format; this eliminates the ambiguity found on other controllers where the duration of the holiday (in days) is specified. A start date may also be set.

Each of the three zones has its own separate self-learning Optimizer. This will advance the heating switch-on time such that the specified temperature is reached at the specified time. The Optimizer uses an outside temperature sensor, and, over a number of days, it learns how each heated zone responds to the heating for different values of outside temperature. The 802 uses this information to construct an internal optimizer advance curve whose shape is continually adjusted. This curve may be inspected and, if necessary, modified. The Optimizer may also be disabled in each zone.

The 802 controller system is supplied in two parts: the computer unit and the interface unit. The computer unit houses the keypad, display and control electronics. The interface unit houses the relays and all connections to the heating system, including sensors. The two units are interconnected with a length (up to 50m) of multicore cable. The interface unit directly switches small 240V AC loads without the need for external relays. Sensor leads may be extended up to 50m.

Ancillary controls

Sunvic Clockbox and Clockbox 2

The Clockbox is a complete plug-in central heating control pack with programming facility into which the motorized valve and cylinder stat are plugged in to. The boiler, pump, room stat and mains are wired in to their respective terminals.

Figure 10.18   *Clockbox – basic programmer*

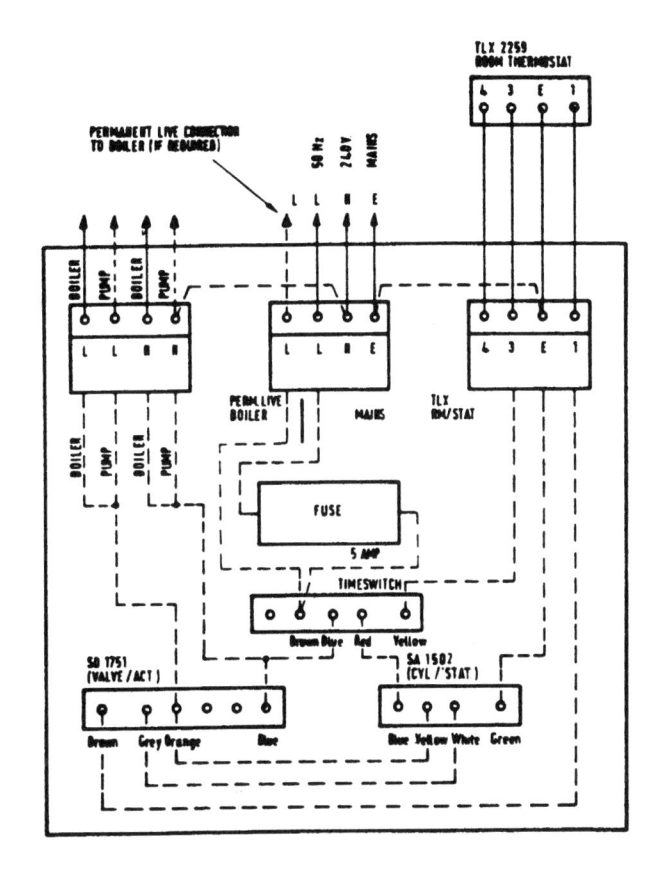

Figure 10.19   *Clockbox 2 – full programmer*

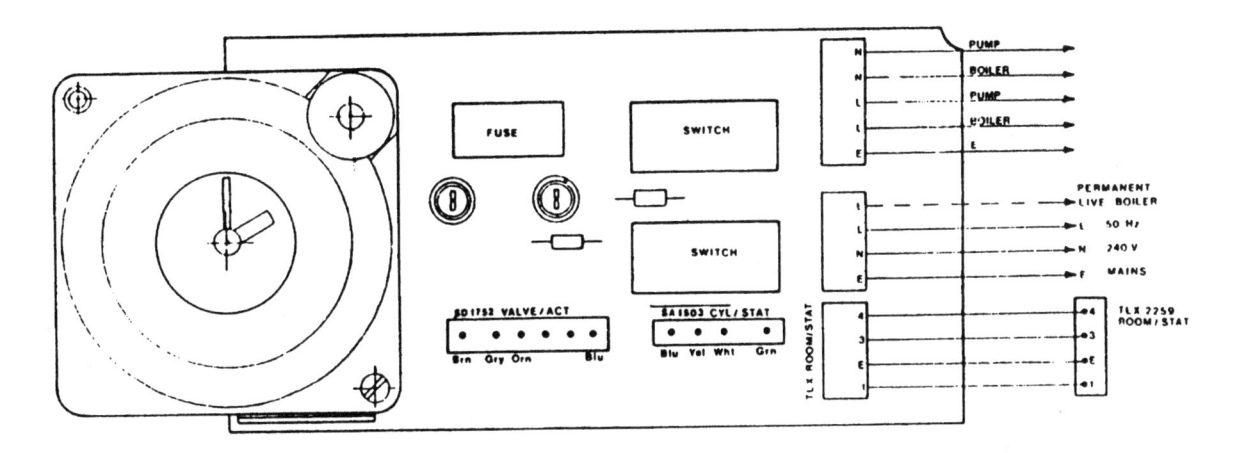

11

Wiring system diagrams

Contents

Gravity hot water, pumped central heating

Figure 11.1	Usual arrangement with room thermostat controlling pump
Figure 11.2	With a cylinder thermostat to control hot water temperature
Figure 11.3	With room and cylinder thermostats controlling boiler and pump giving hot water priority
Figure 11.4	With room and cylinder thermostats both controlling pump, giving hot water priority
Figure 11.5	With pump on each circuit, giving limited hot water temperature control
Figure 11.6	With one 2-port spring return motorized valve in central heating circuit, giving limited temperature control of hot water
Figure 11.7	With a 28mm 2-port spring return motorized valve to hot water circuit, giving full temperature and programming control
Figure 11.8	With a 2-port spring return motorized valve to both central heating and hot water circuits, giving full temperature control
Figure 11.9	With a 2-port spring return motorized valve to hot water circuit and relay, giving full temperature and programming control
Figure 11.10	With a 2-port motor open, motor close motorized valve to both central heating and hot water circuits, giving full temperature control
Figure 11.11	With a 2-port motor open, motor closed motorized valve to hot water circuit and relay, giving full temperature and programming control
Figure 11.12	Danfoss Plan 1.1 and Plan 2A
Figure 11.13	Utilizing Drayton SU1 switch unit
Figure 11.14	Utilizing Drayton SU2 switch unit
Figure 11.15	With Sunvic SZ 1302 actuator to hot water circuit, giving full temperature and programming control

Fully pumped systems

Figure 11.16	For use on one circuit only, i.e. central heating or hot water, with or without temperature control
Figure 11.17	Priority system with diverter valve showing hot water priority
Figure 11.18	Priority system with diverter valve showing central heating priority
Figure 11.19	Priority system with diverter valve and simple changeover switch to provide for optional priority
Figure 11.20	With one 2-port motor open, motor close motorized valve in central heating circuit
Figure 11.21	With one 2-port spring return motorized valve in central heating circuit, providing limited hot water temperature control
Figure 11.22	With a 2-port spring return motorized valve to both central heating and hot water circuits, providing full temperature and programming control

Wiring system diagrams

Figure 11.23   With a mid-position 3-port valve with standard colour flex conductors. Provides full temperature and programming control

Figure 11.24   With a 2-port motor open, motor close motorized valve to both central heating and hot water circuits. Provides full temperature and programming control

Figure 11.25   With a 2-port motor open, motor close motorized valve and a 2-port spring return motorized valve to control central heating and hot water circuits

Figure 11.26   With a 2-port spring return motorized valve to both central heating and hot water circuit when no permanent live is available for valve auxiliary switches. Provides full temperature control

Figure 11.27   With a 2-port spring return *normally open* motorized valve to both central heating and hot water circuits

Figure 11.28   Showing the addition of a 2-port spring return motorized valve to an existing system for zone control

Figure 11.29   Showing multiples of 2-port spring return motorized valves wired to one programmer, as may be used in a large property for zone control

Figure 11.30   Showing multiples of 2-port motor open, motor close motorized valves wired to one programmer, as may be used in a large property for zone control

Figure 11.31   With one boiler and zone control using one pump for each zone. No motorized valves

Figure 11.32   With two or more pump overrun boilers and one pump using relays as required

Figure 11.33   With pump to each circuit and no motorized valves

Figure 11.34   With pump overrun boiler, pump to each circuit, room and cylinder thermostats, programmer and relay. No motorized valves

Figure 11.35   With pump overrun boiler, pump and 2-port spring return motorized valves to each circuit, room and cylinder thermostats, programmer and relay

Figure 11.36   ACL Biflo System MK1 with time clock

Figure 11.37   ACL Biflo System MK1 using a programmer with voltage free contacts. Provides full temperature and programming control

Figure 11.38   Danfoss 2.2 and 2C System

Figure 11.39   Drayton Plan 1 with Drayton TA/M2 actuator

Figure 11.40   Drayton Flowshare 5 System with TA/M4 actuator, RB1 relay box and programmer

Figure 11.41   Drayton Flowshare 5 System with Drayton TA/M4 actuator, RB2 relay box and programmer

Figure 11.42   Drayton Plan 7 System with two TA/M2A actuators providing full temperature and programming control

Figure 11.43   Homewarm Manual System

Figure 11.44   Homewarm Auto System with Switchmaster VM5 actuator

Figure 11.45   Honeywell Y Plan System using the V4073 6-wire, 3-port motorized valve with integral relay giving full temperature control

Figure 11.46   Honeywell Y Plan System using the V4073 6-wire, 3-port motorized valve with integral relay giving full temperature and programming control

Figure 11.47   Landis & Gyr LGM System

Figure 11.48   SMC Control Pack 2 System with one boiler and a pump to both central heating and hot water circuits. No motorized valves. Provides full temperature and programming control

Figure 11.49   SMC Control Pack 2 System with pump overrun boiler and a pump to both central heating and hot water circuits. No motorized valves. Provides full temperature and programming control

Figure 11.50   Sunvic Duoflow System with RJ 1801 relay box and programmer

Figure 11.51   Sunvic Duoflow System with RJ 2801 relay box and programmer

Figure 11.52   Sunvic Duoflow System with RJ 2802 or RJ 2852 relay box and time clock

Figure 11.53   Sunvic Duoflow System with RJ 2802 or RJ 2852 relay box and programmer

Figure 11.54   Switchmaster Midi System

183

Various frost protection thermostat wiring

Figure 11.55 Wiring of a double pole frost thermostat to a gravity hot water, pumped central heating system
Figure 11.56 Wiring of a single pole frost thermostat to a gravity hot water, pumped central heating system using a DPDT relay
Figure 11.57 Wiring a frost thermostat to a motor open, motor close motorized valve
Figure 11.58 Wiring a frost thermostat to a fully pumped system using two motor open, motor close motorized valves

Supplementary wiring diagrams

Figure 11.59 Utilizing a full control programmer for basic control
Figure 11.60 Addition of a changeover switch to a priority system incorporating a 3-port valve, e.g. Honeywell V4044, to enable changing from hot water or heating priority as required
Figure 11.61 Simple pump overrun thermostat wiring
Figure 11.62 Wiring of relays, including examples
Figure 11.63 Wiring of a relay to allow a 2-wire (SPST) cylinder or room thermostat to function as a 3-wire (SPDT)
Figure 11.64 Simplified warm air unit wiring

Gravity hot water, pumped central heating

Figure 11.1 *Usual arrangement with room thermostat controlling pump*

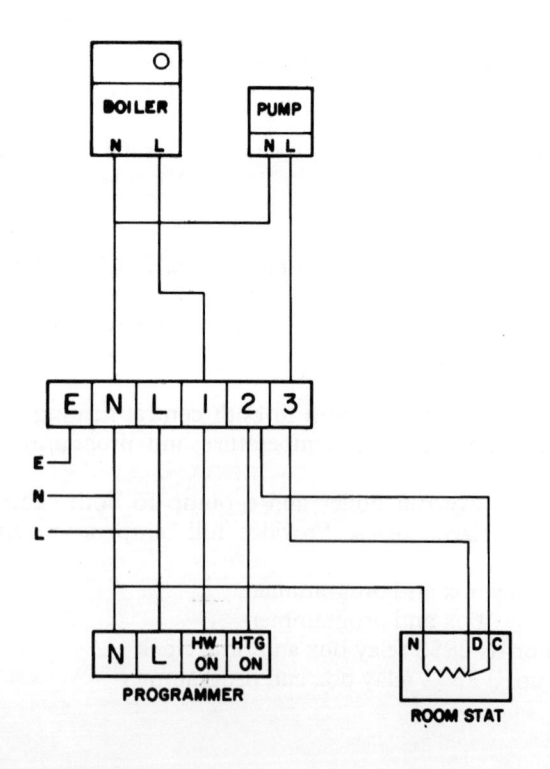

Notes
a) Known as Drayton Plan 2.
b) Suitable only for Basic programming.
c) Connect frost thermostat across junction box terminals L–1 for hot water or double pole frost thermostat L–1 and L–3.

Wiring system diagrams

Figure 11.2   *With a cylinder thermostat to control hot water temperature*

Notes
a) Known as Honeywell 'A' Plan.
b) Suitable only for Basic programming.
c) Cylinder thermostat will only control hot water temperature when 'Hot Water Only' is selected on programmer.
d) Connect frost thermostat across junction box terminals L–4 for hot water or use double pole frost thermostat L–4 and L–3.

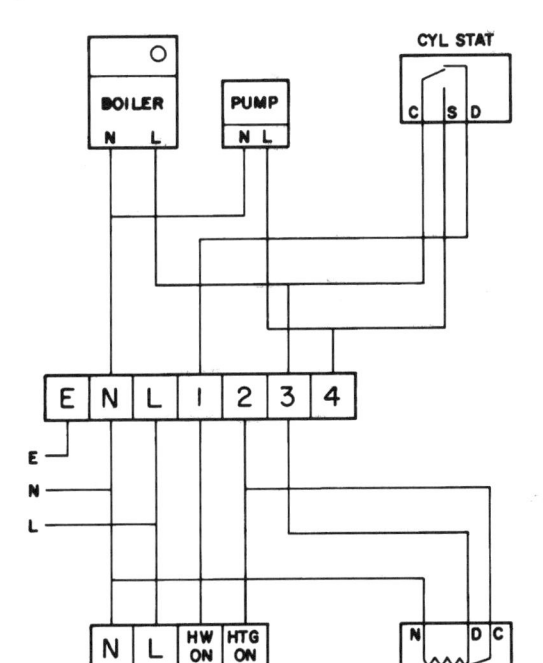

Figure 11.3   *With room and cylinder thermostats controlling boiler and pump giving hot water priority*

Notes
a) Suitable for Basic or Full control programming but no temperature control of hot water when heating is on.
b) Heating will not work until cylinder thermostat is satisfied.
c) Connect frost thermostat across junction box terminals L–1.

185

Installing and Servicing Domestic Heating Wiring Systems and Control

Figure 11.4 *With room and cylinder thermostats both controlling pump, giving hot water priority*

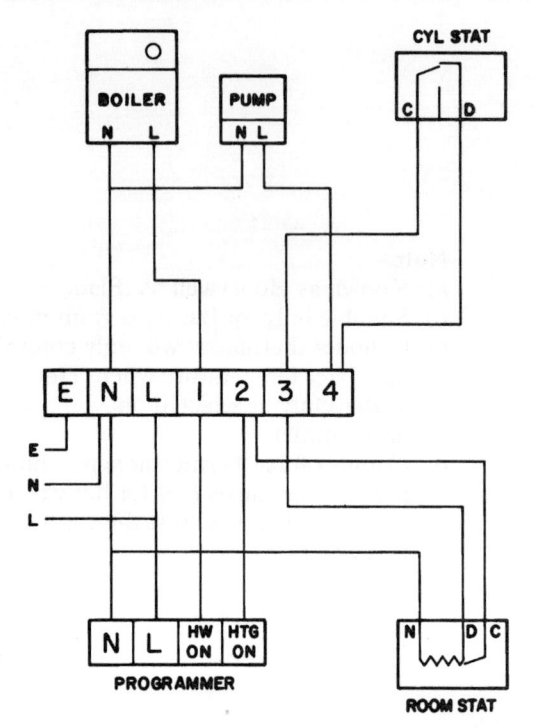

Notes
a) This method provides for the heating of the hot water to the cylinder thermostat temperature before the heating pump can function. This is ideal for systems where the heating pump will 'starve' the cylinder.
b) No hot water temperature control. Cylinder thermostat serves only to delay heating until cylinder is hot.
c) Suitable for Basic programmers only.
d) Connect frost thermostat L–1 for hot water or double pole frost thermostat L–1 and L–4.

Figure 11.5 *With pump on each circuit, giving limited hot water temperature control*

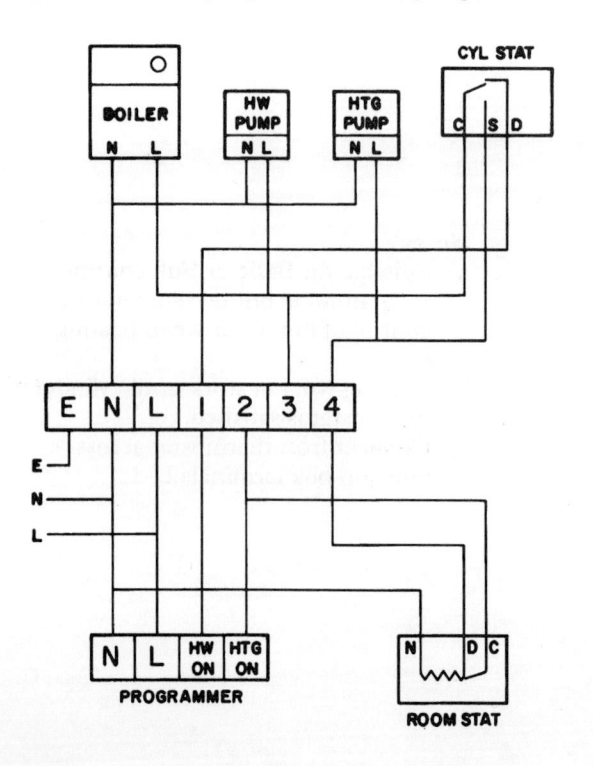

Notes
a) Suitable only for Basic programming.
b) Hot water temperature control only possible when 'Hot Water Only' selected.
c) Connect frost thermostat across junction box terminals L–3 for hot water only or double pole frost thermostat L–3 and L–4.

Wiring system diagrams

Figure 11.6 *With one 2-port spring return motorized valve in central heating circuit, giving limited temperature control of hot water*

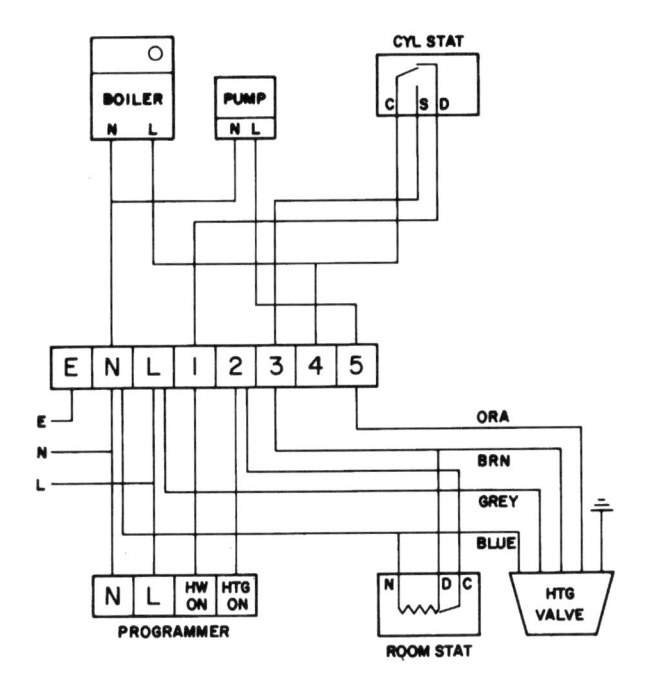

Notes
a) Known as Honeywell System L.
b) Suitable only for Basic programming.
c) Hot water temperature control only possible when 'Hot Water Only' selected.
d) Connect frost thermostat across junction box terminals L–3.

Figure 11.7 *With a 28mm 2-port spring return motorized valve to hot water circuit, giving full temperature and programming control*

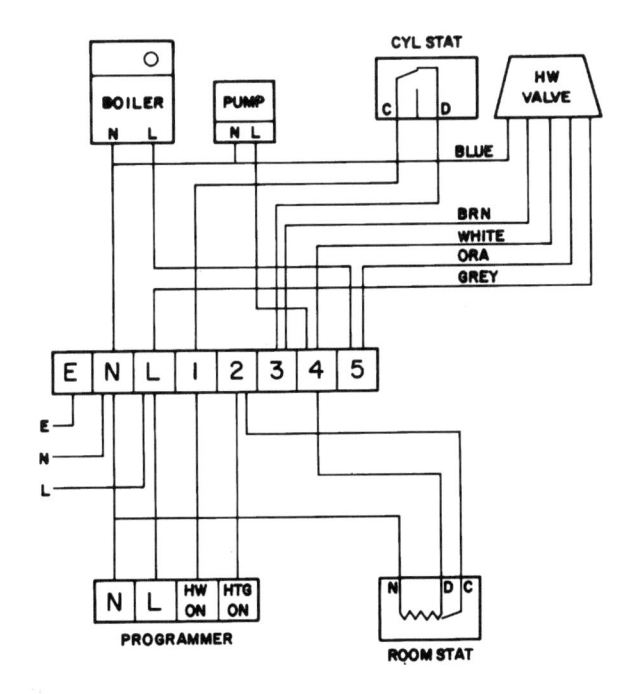

Notes
a) Known as Honeywell 'C' Plan.
b) Suitable for Basic or Full programming.
c) Connection of grey can be made to junction box terminal 1 (Hot Water On) if no live at junction box.
d) Motorized valve must have changeover auxiliary switch.

Installing and Servicing Domestic Heating Wiring Systems and Control

Figure 11.8 *With a 2-sport spring return motorized valve to both central heating and hot water circuits, giving full temperature control*

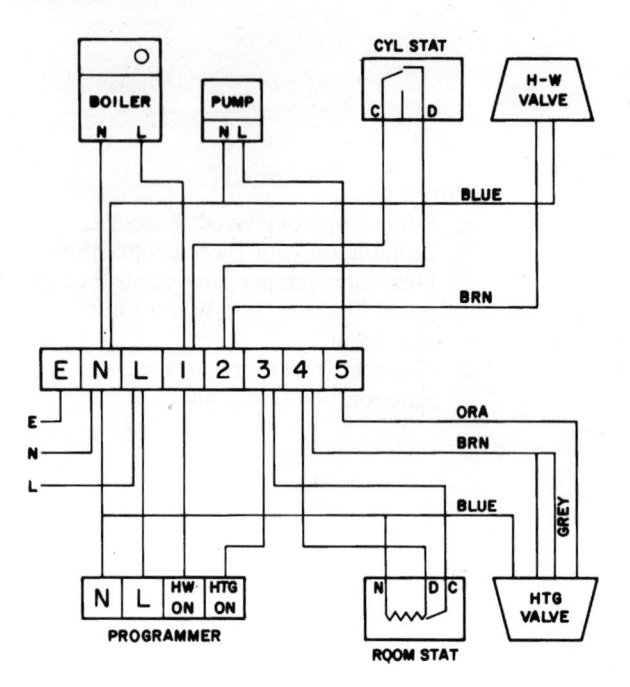

Notes
a) Suitable only for Basic programming.
b) Connect frost thermostat across junction box terminals L–1 for hot water only or double pole frost thermostat L–1 and L–4.

Figure 11.9 *With a 2-port spring return motorized valve to hot water circuit and relay, giving full temperature and programming control*

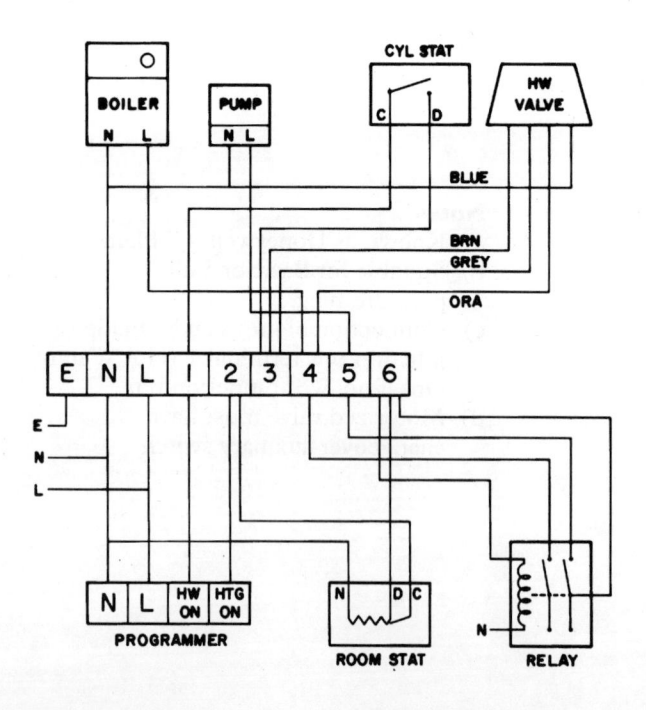

Notes
a) Suitable for Basic or Full programming.
b) Requires a DPDT relay.
c) Connect frost thermostat across junction box terminals L–6.

Wiring system diagrams

Figure 11.10 *With a 2-port motor open, motor close motorized valve to both central heating and hot water circuits, giving full temperature control*

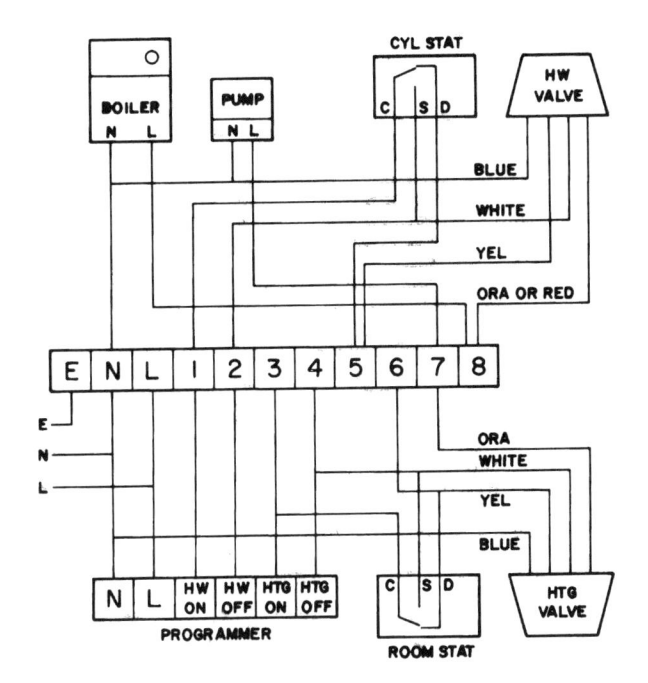

Notes

a) Motorized valve on Heating to stop gravity circulation.
b) System requires changeover room and cylinder thermostats.
c) Suitable for Basic control only but programmer must have 'Offs' to heating and hot water.
d) For frost protection see special diagram, page 212

Figure 11.11 *With a 2-port motor open, motor close motorized valve to hot water circuit and relay, giving full temperature and programming control*

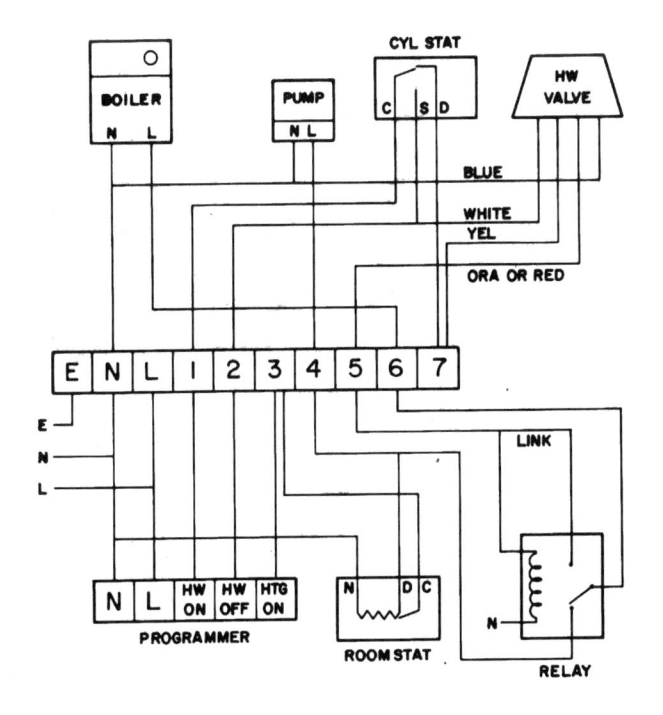

Notes

a) Suitable for Basic or Full programming but 'Hot Water Off' signal required regardless of whichever is used.
b) Requires an SPDT relay.
c) Connect frost stat across junction box terminals L–4.

Installing and Servicing Domestic Heating Wiring Systems and Control

Figure 11.12 *Danfoss Plan 1.1 and Plan 2A*

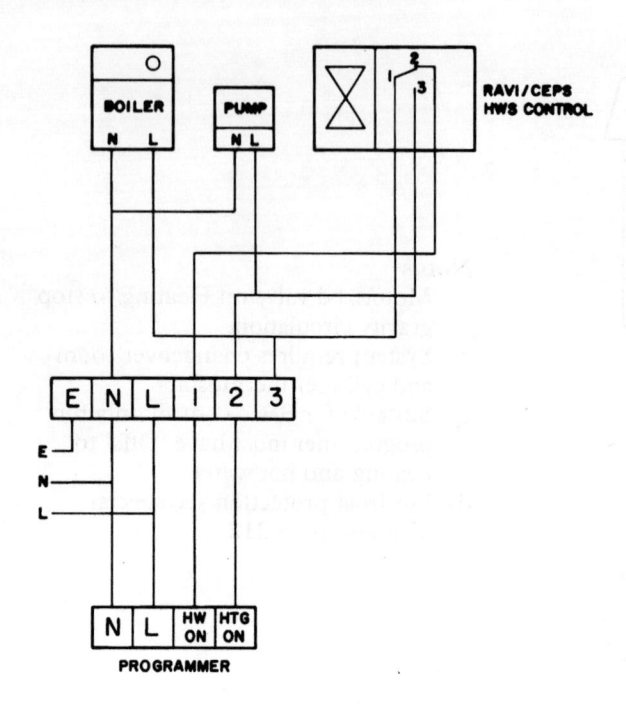

Notes
a) Suitable only for Basic programming.
b) Connect frost thermostat across junction box terminals L–3 for hot water or double pole frost thermostat L–3 and L–2.

Figure 11.13 *Utilizing Drayton SU1 switch unit*

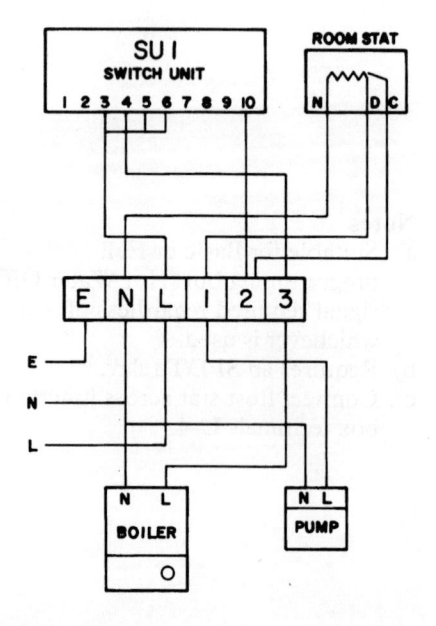

Notes
a) Could be used with time clock.
b) Connect frost thermostat across junction box terminals L–3 for hot water or double pole frost thermostat L–3 and L–1.

Wiring system diagrams

Figure 11.14 *Utilizing Drayton SU2 switch unit*

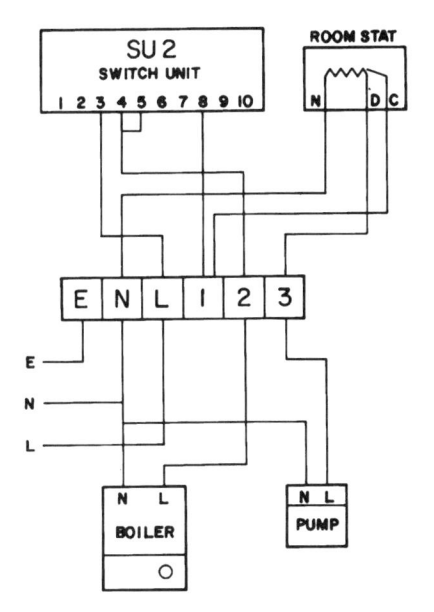

Notes
a) Could be used with time clock.
b) Connect frost thermostat across junction box terminals L–2 for hot water or double pole frost thermostat L–2 and L–3.

Figure 11.15 *With Sunvic SZ1302 actuator to hot water circuit, giving full temperature and programming control*

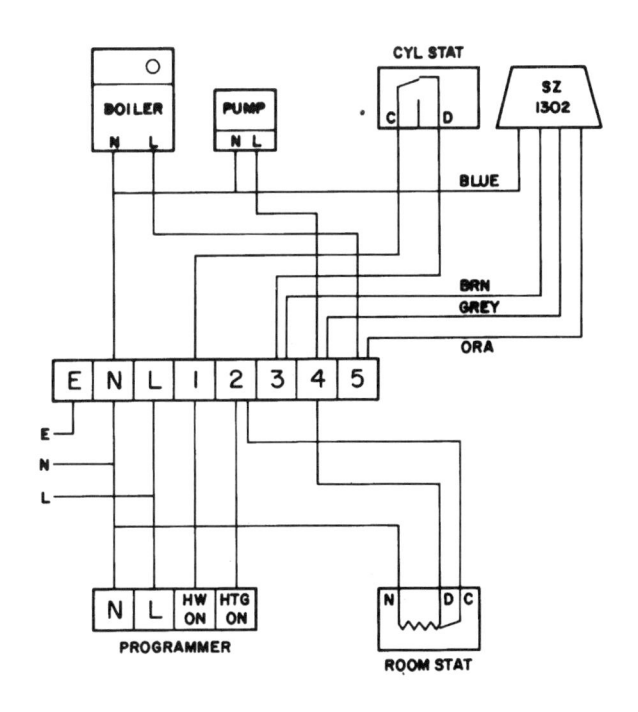

Notes
a) Suitable for Basic or Full programming.
b) Connect frost thermostat across junction box terminals L–4.

191

Installing and Servicing Domestic Heating Wiring Systems and Control

Fully pumped systems

Figure 11.16 *For use on one circuit only, i.e. central heating or hot water, with or without temperature control*

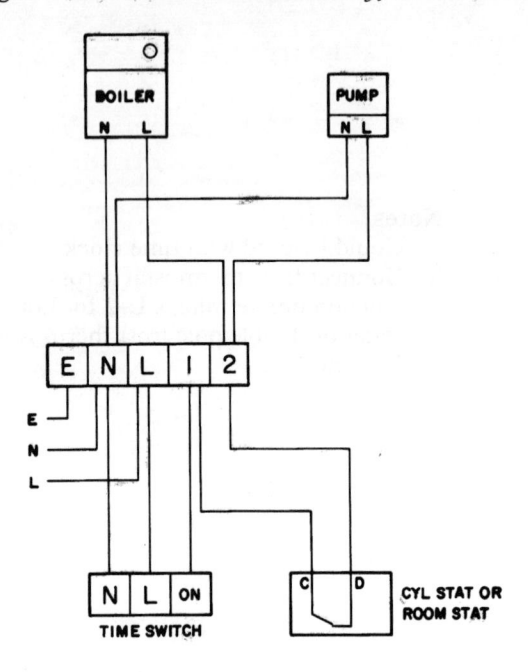

Notes
a) Where no room/cylinder thermostat exists, link 1–2 in junction box.
b) When using a pump overrun boiler then wire pump as indicated in Chapter 8
c) Connect frost thermostat across junction box terminals L–1.

Figure 11.17 *Priority system with diverter valve showing hot water priority*

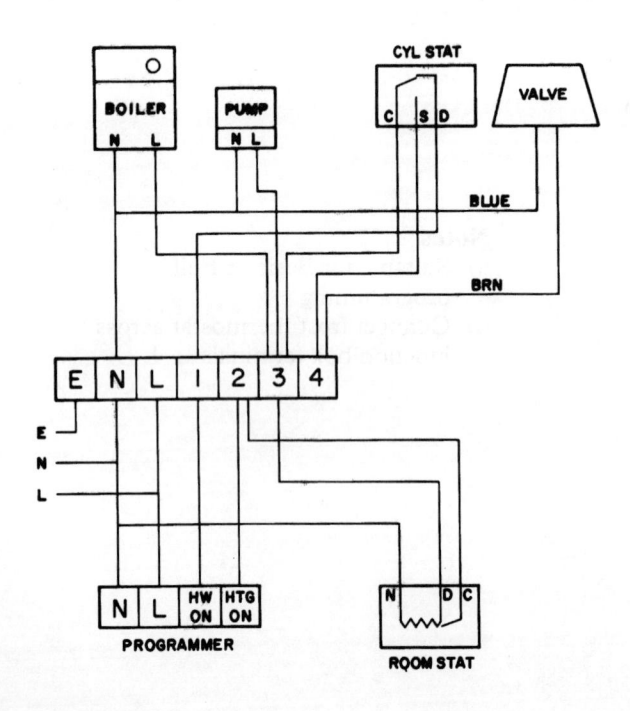

Notes
a) Suitable for Basic programmer only.
b) For specially manufactured priority programmers refer to page 58.
c) Connect frost thermostat across junction box terminals L–3.

Wiring system diagrams

Figure 11.18 *Priority system with diverter valve showing central heating priority*

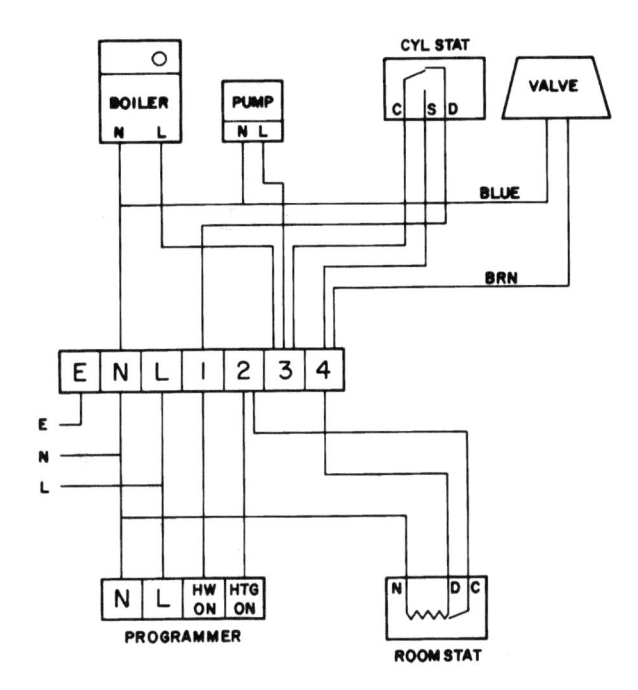

Notes
a) Suitable only for Basic programming.
b) For specially manufactured priority programmers refer to programmer section, page 58
c) Connect frost thermostat across junction box terminals L–3.

Figure 11.19 *Priority system with diverter valve and simple changeover switch to provide for optional priority*

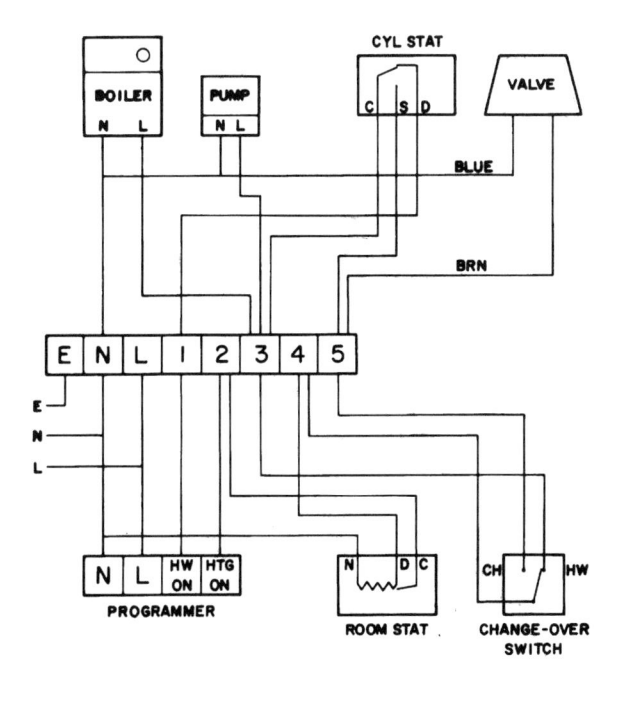

Notes
a) Suitable for Basic programming only.
b) Connect frost thermostat across junction box terminals L–3.

193

Installing and Servicing Domestic Heating Wiring Systems and Control

Figure 11.20   With one 2-port motor open, motor close motorized valve in central heating circuit

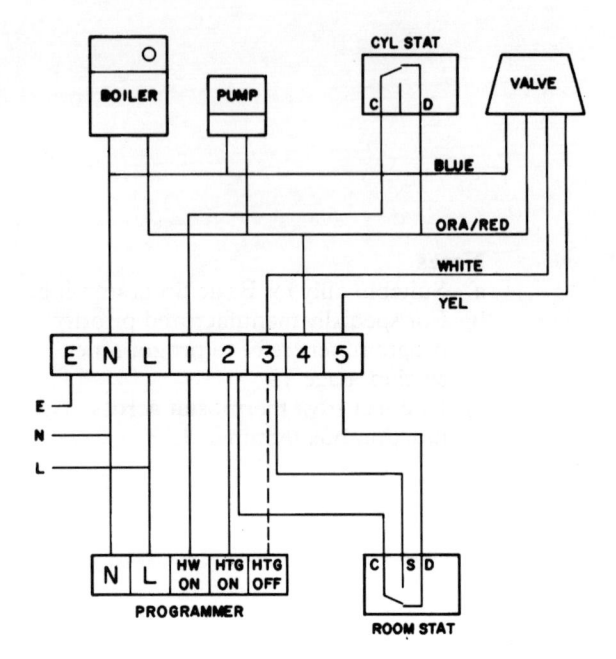

Notes

a) Hot water temperature control only when programmer is in 'Hot Water Only' position.
b) Programmer should have 'Heating Off' signal, otherwise room thermostat will need to be turned down whilst clock is in 'Heating On' mode.
c) System requires changeover room thermostat.
d) Suitable for Basic control only.
e) Connect frost thermostat across junction box terminals L–5.

Figure 11.21   With one 2-port spring return motorized valve in central heating circuit, providing limited hot water temperature control

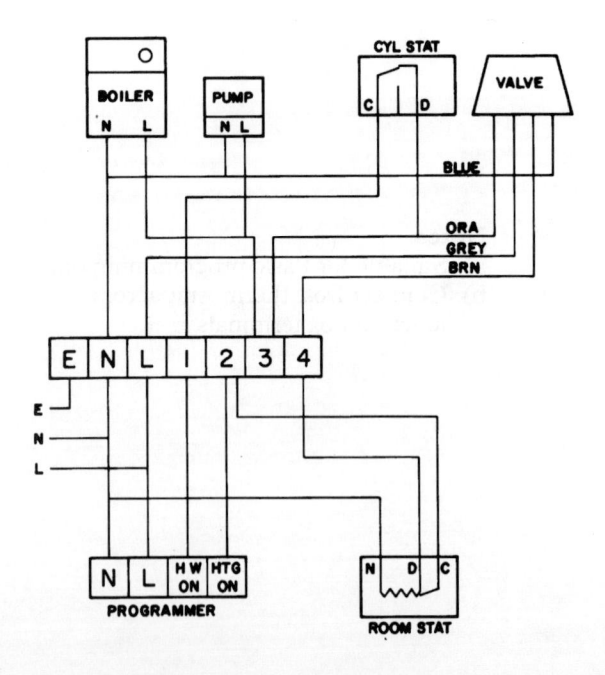

Notes

a) Suitable for Basic programming only.
b) Hot water temperature control only when programmer set to 'Hot Water' only.
c) Connect frost thermostat across junction box terminals L–4.

Wiring system diagrams

Figure 11.22 *With a 2-port spring return motorized valve to both central heating and hot water circuits, providing full temperature and programming control*

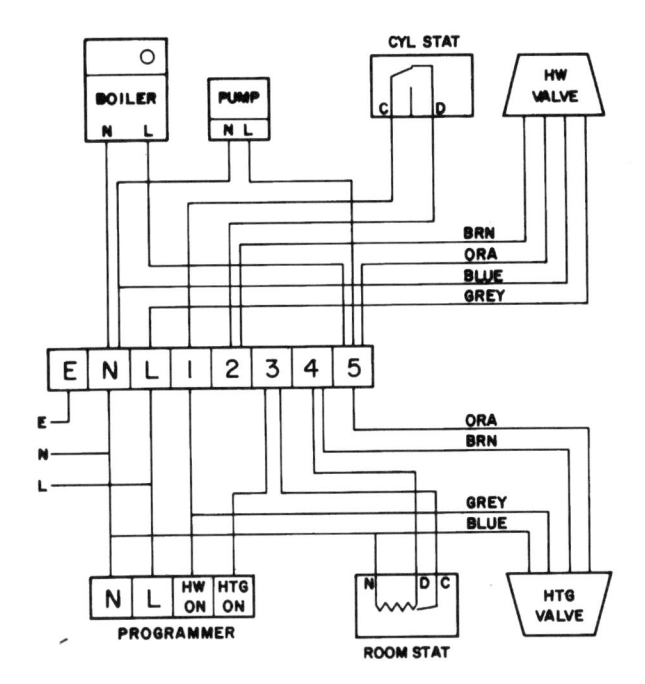

Notes
a) Known as Honeywell S Plan and Sunvic System 4.
b) Suitable for Basic or Full programming.
c) Connect frost thermostat across junction box terminals L–4.

Figure 11.23 *With a mid-position 3-port valve with standard colour flex conductors. Provides full temperature and programming control*

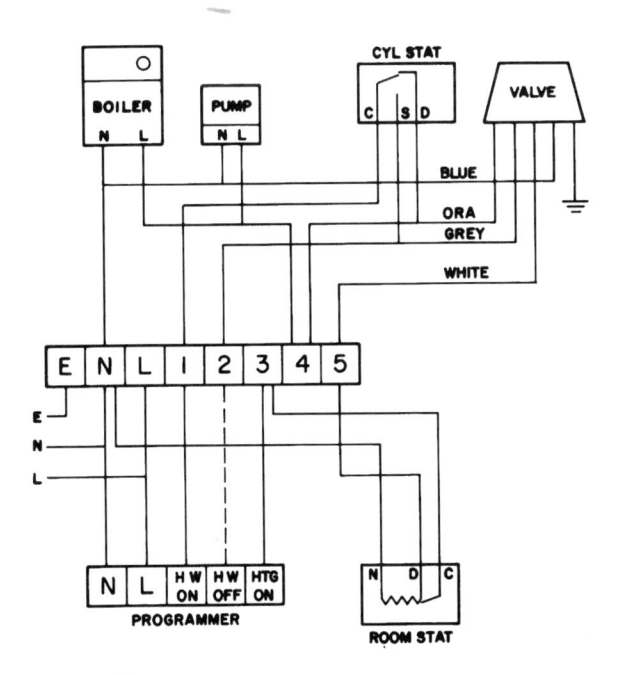

Notes
a) Full programming only available if programmer has 'Hot Water Off' terminal and dotted connection is made.
b) Known as Honeywell Y Plan and Sunvic Unishare System.
c) Connect frost thermostat across junction box terminals L–5.

Installing and Servicing Domestic Heating Wiring Systems and Control

Figure 11.24 With a 2-port motor open, motor close motorized valve to both central heating and hot water circuits. Provides full temperature and programming control

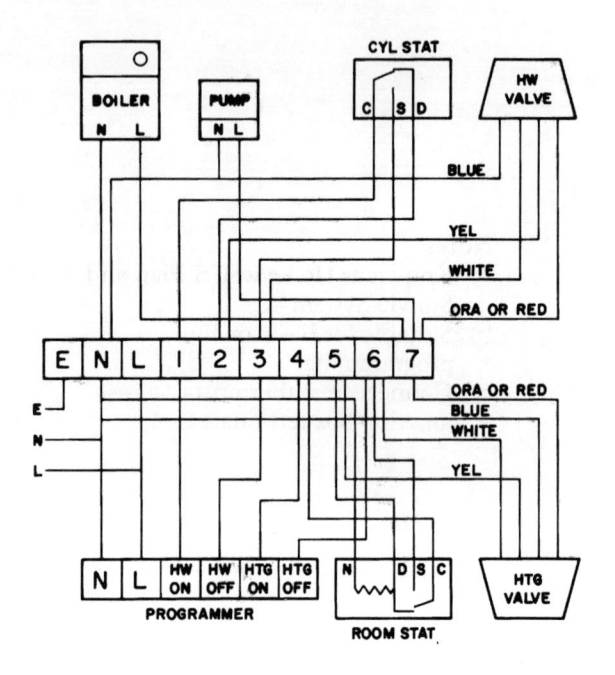

Notes
a) Programmer must have 'Offs' for both heating and hot water.
b) For frost protection refer to special diagrams.

Figure 11.25 With a 2-port motor open, motor close motorized valve and a 2-port spring return motorized valve to control central heating and hot water circuits

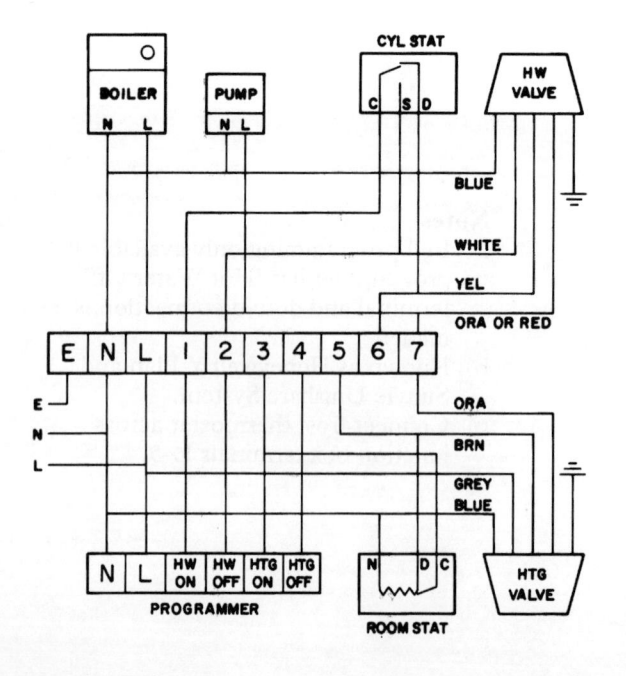

Notes
a) Suitable for Basic or Full control but programmer must have 'Off' signal for circuit with motor open, motor close valve (shown on hot water in this diagram).
b) Connect frost thermostat across junction box terminals L–5.

Wiring system diagrams

Figure 11.26 With a 2-port spring return motorized valve to both central heating and hot water circuit when no permanent live is available for valve auxiliary switches. Provides full temperature control

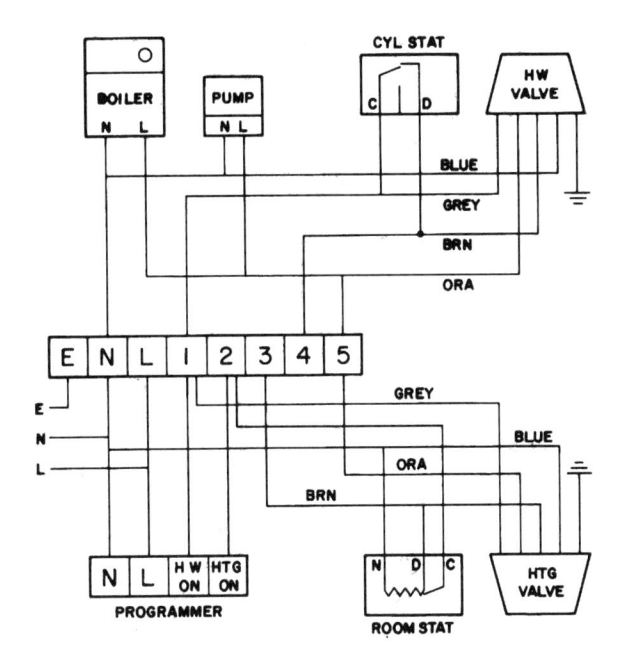

Notes
a) Suitable for Basic programming only.
b) Connect frost thermostat across junction box terminals L–3.

Figure 11.27 With a 2-port spring return **normally open** motorized valve to both central heating and hot water circuits

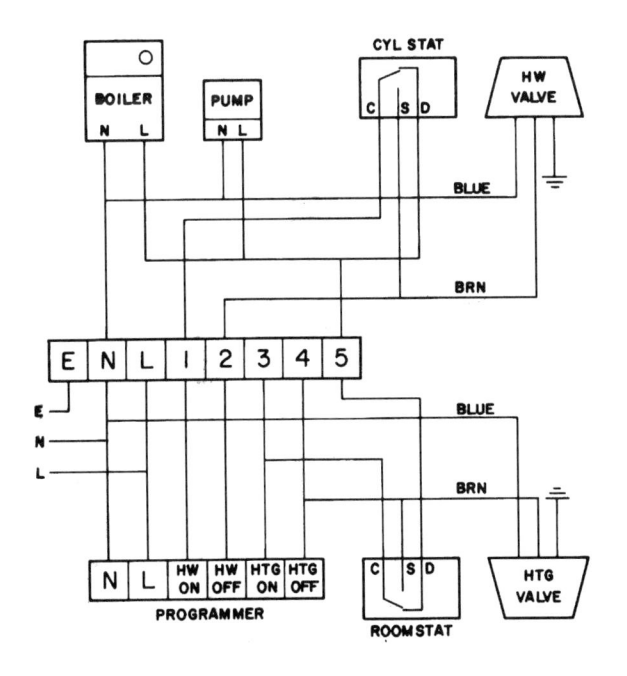

Notes
a) Known as Honeywell System G.
b) Not commonly used but type of valve V4043B used in solid fuel systems.
c) Essential that programmer has 'Off' signals to both heating and hot water.
d) System requires changeover room and cylinder thermostat.
e) Connect frost thermostat across junction box terminals L–5.

Installing and Servicing Domestic Heating Wiring Systems and Control

Figure 11.28 *Showing the addition of a 2-port spring return motorized valve to an existing system for zone control*

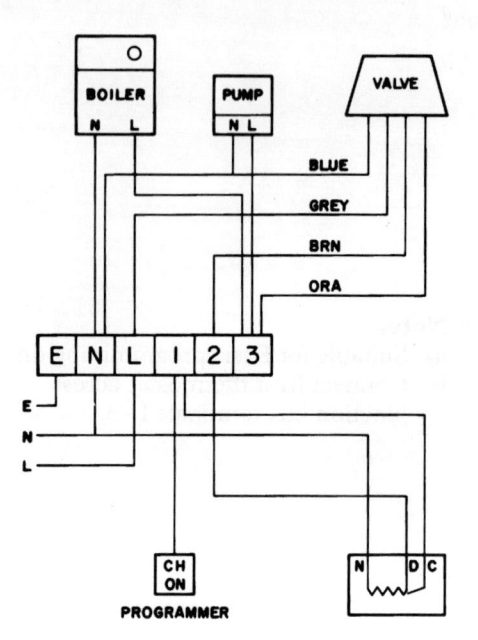

Notes
Shown on central heating but could apply to the hot water circuit if required.

Figure 11.29 *Showing multiples of 2-port spring return motorized valves wired to one programmer, as may be used in a large property for zone control*

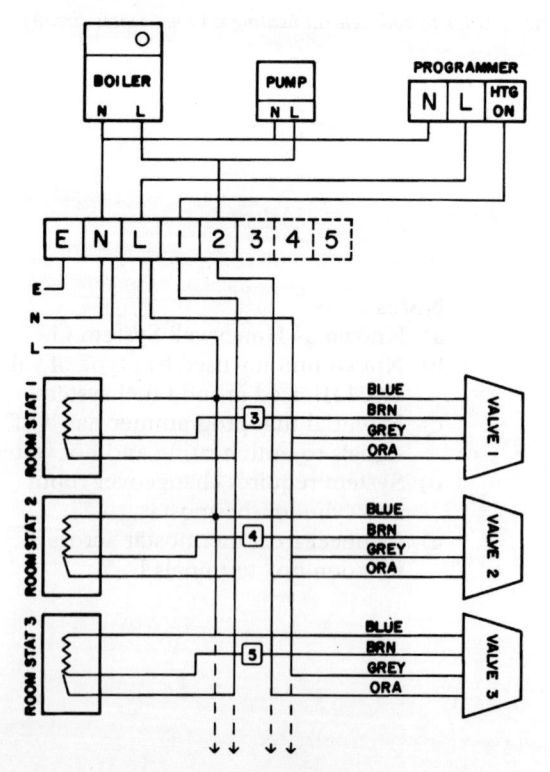

Notes
Up to three valves are shown on heating, although in theory there is no limit and they could also be used in the hot water circuit, where several cylinders are utilized.

198

Wiring system diagrams

Figure 11.30 *Showing multiples of 2-port motor open, motor close motorized valves wired to one programmer, as may be used in a large property for zone control*

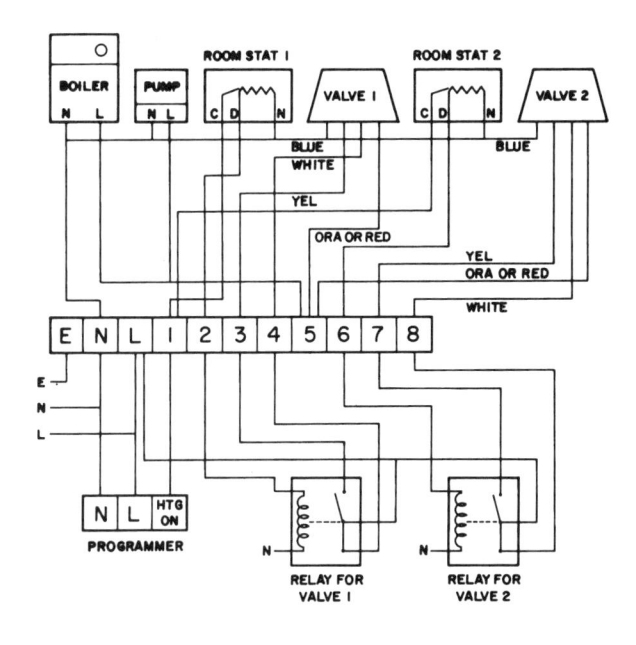

Notes

a) Diagram shows two valves on the heating, although in theory any number can be used, providing a relay is used for each. Note that the 'Heating Off' signal from the programmer is not required. The same principle applies if used on the hot water circuit.

b) In a system utilizing, for example, one motor open, motor close valve on the hot water and two motor open, motor close valves on the heating, the hot water valve would be wired as normal using the 'Hot Water Off' from programmer and cylinder thermostat, and the heating valves should be wired as above. However, if there were two motor open, motor close valves to both hot water and heating circuits then all four would need to be wired as shown, and four relays would be required. As this would normally occur in a large property, it would be easier and more cost effective, and provide greater benefits in programming to fit two programmers or change actuators to spring return type. Remember that where a valve is controlled by a relay, a changeover thermostat or 'Off' signal from the programmer is not required.

Figure 11.31 *With one boiler and zone control using one pump for each zone. No motorized valves*

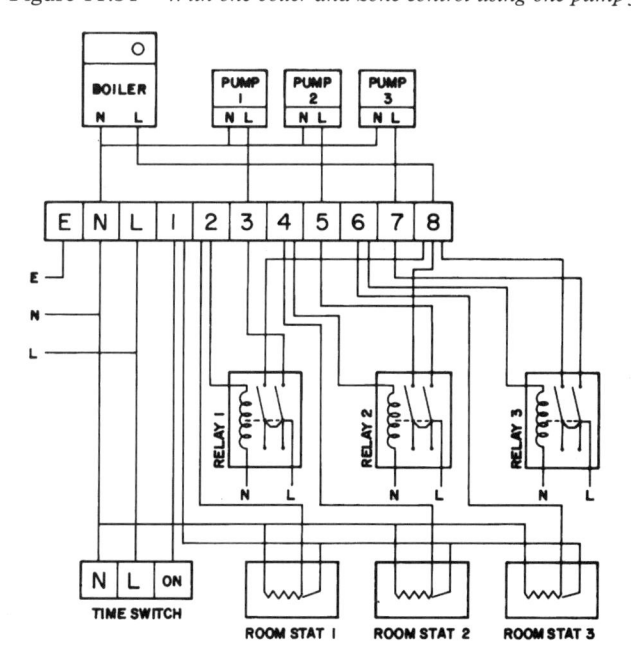

Notes

a) Requires a DPDT relay for each pump.

b) More appropriate to a commercial or large domestic situation.

c) Connect frost thermostat across junction box terminals L–1. It is assumed that at least one room thermostat is calling for heat.

Installing and Servicing Domestic Heating Wiring Systems and Control

Figure 11.32 *With two or more pump overrun boilers and one pump using relays as required*

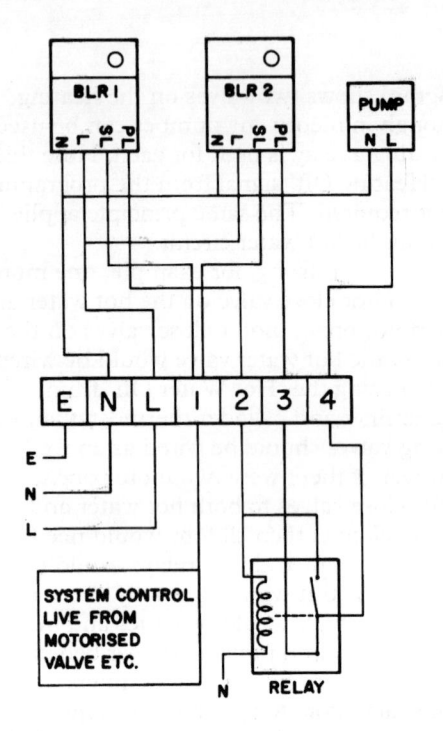

Notes
a) Requires a single pole changeover relay for each boiler after the first one, e.g. two boilers, one relay, three boilers, 2 relays, etc.
b) Boiler PL refers to pump live. Boiler SL refers to switched live from system control.
c) Connect frost thermostat across junction box terminals L–2.

Figure 11.33 *With pump to each circuit and no motorized valves*

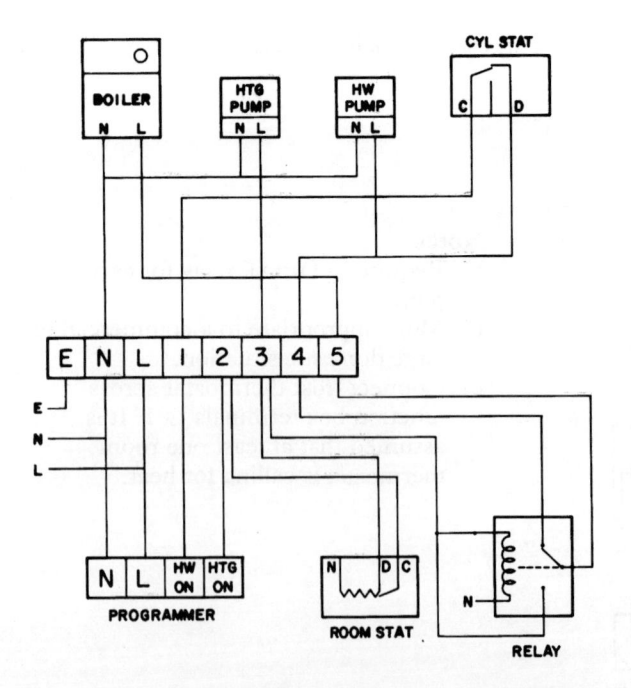

Notes
a) Requires a single pole changeover relay (SPDT).
b) System similar to SMC control pack MK1 and can be utilized to convert from original Horstmann programmer.
c) Basic or Full control, dependent on programmer used.
d) Connect frost thermostat across junction box terminals L–3.

Wiring system diagrams

Figure 11.34 *With pump overrun boiler, pump to each circuit, room and cylinder thermostats, programmer and relay. No motorized valves*

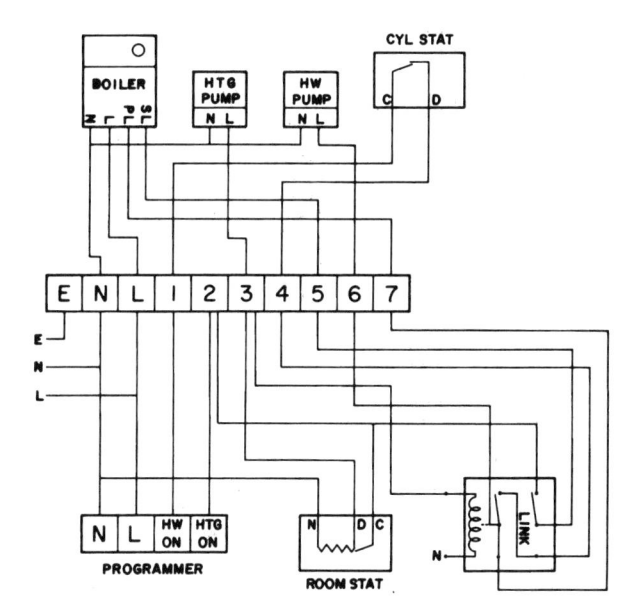

Notes
a) Requires a double pole changeover relay (DPDT).
b) System similar to SMC control pack MK1 and can be utilized to convert from original Horstmann programmer.
c) Basic or Full control dependent on programmer used.
d) Connect frost thermostat across junction box terminals L–3.

Figure 11.35 *With pump overrun boiler, pump and 2-port spring return motorized valve to each circuit, room and cylinder thermostats, programmer and relay*

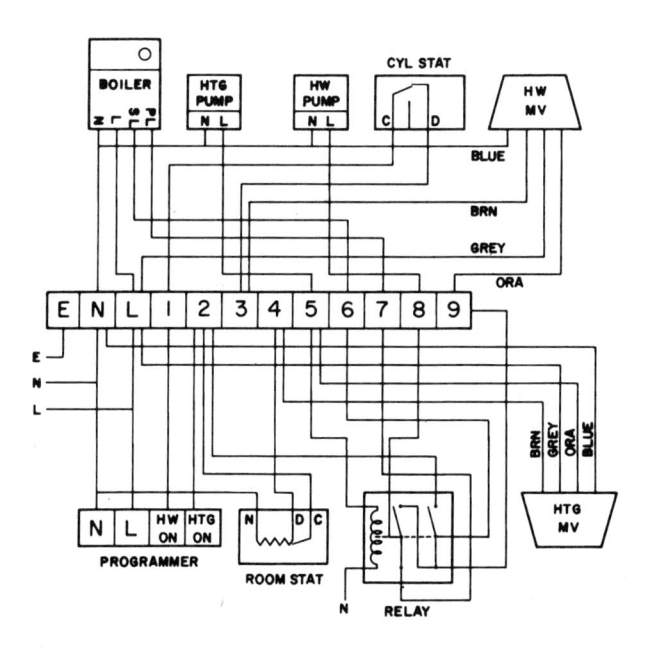

Notes
a) Requires a double pole, changeover relay (DPDT).
b) Pump overrun functions on hot water pump.
c) Connect frost thermostat across junction box terminals L–4.

201

Installing and Servicing Domestic Heating Wiring Systems and Control

Figure 11.36   *ACL Biflo system MK1 with time clock*

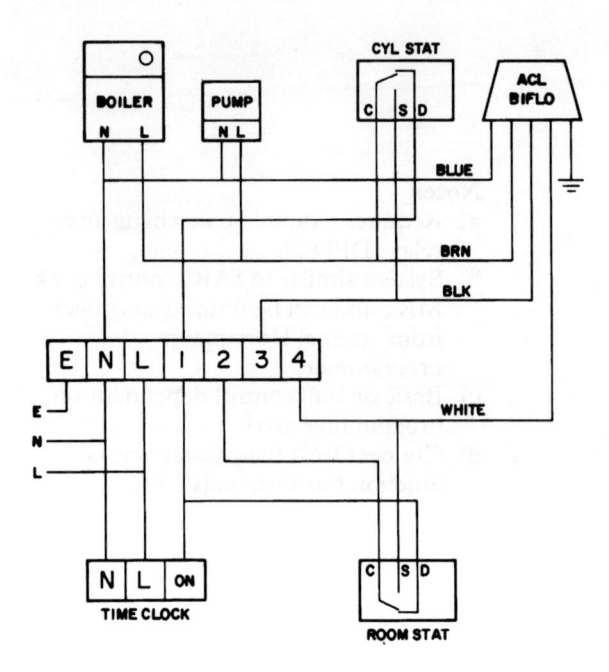

Notes
a) System requires changeover room and cylinder thermostats.
b) See motorized valve section (Chapter 6, ACL 672 BRO 340).
c) Connect frost thermostat across junction box terminals L–1.

Figure 11.37   *ACL Biflo system MK1 using a programmer with voltage free contacts. Provides full temperature and programming control*

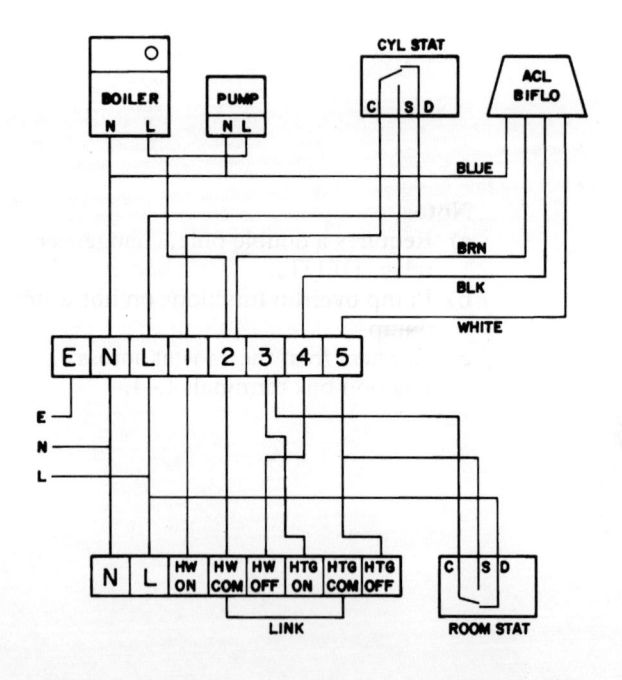

Notes
a) Programmer must have voltage free contacts – link as shown.
b) Brown wire of valve may be red.
c) System requires changeover room and cylinder thermostats.
d) Connect frost thermostat across junction box terminals L–2.

Wiring system diagrams

Figure 11.38 *Danfoss 2.2 and 2C system*

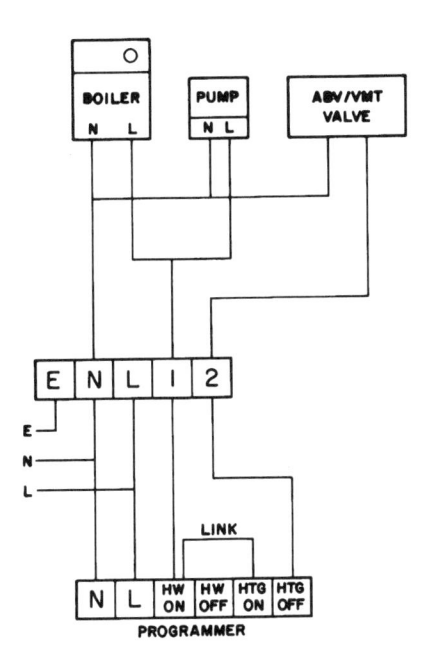

Notes
a) Suitable for Basic programming only.
b) Connect frost thermostat across junction box terminals L–1.

Figure 11.39 *Drayton Plan 1 with Drayton TA/M2 actuator*

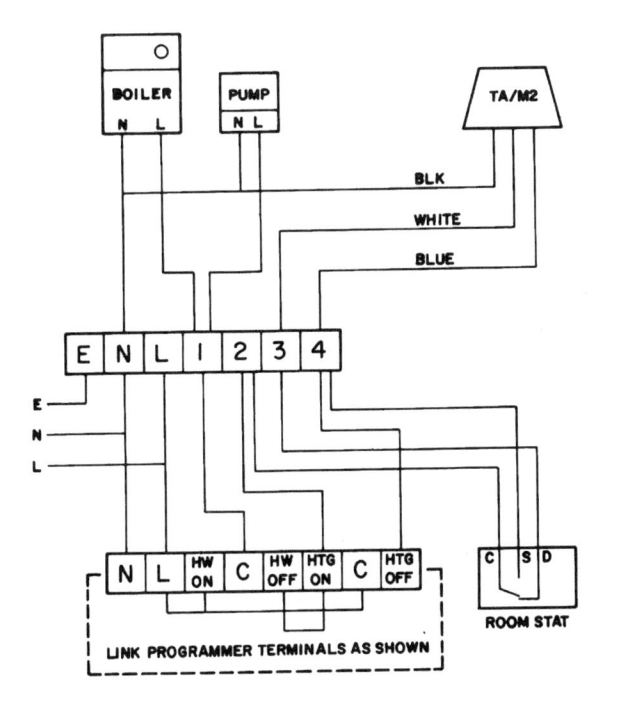

Notes
a) No temperature control of hot water unless non-electric device is used.
b) Suitable for Basic programmer only, although programmer must have hot water and heating 'Offs' and voltage free switching.
c) Note that valve black is neutral.
d) Connect frost thermostat across junction box terminals L–3.

203

Installing and Servicing Domestic Heating Wiring Systems and Control

Figure 11.40 *Drayton Flowshare 5 system with Drayton TA/M4 actuator, RB1 relay box and programmer*

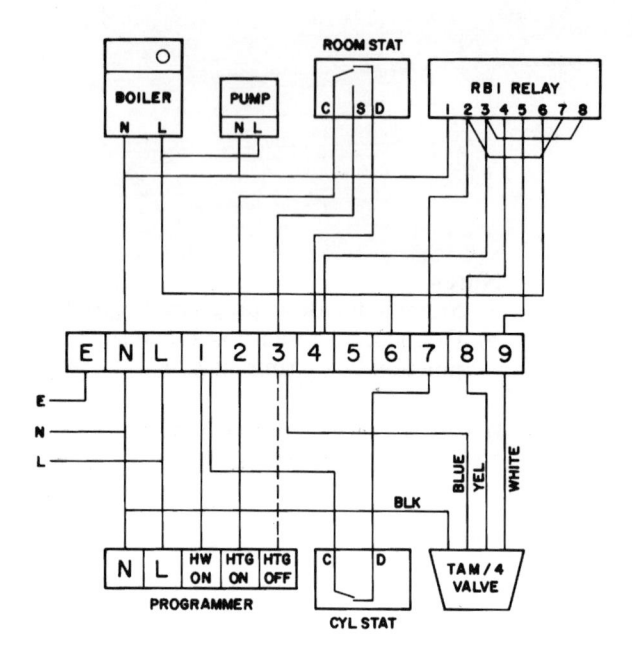

Notes

a) Suitable for Basic programmers only unless programmer has 'Heating Off' facility and dotted line is connected.
b) If a time switch is used link 'Heating On' and 'Hot Water On'.
c) As this actuator is reversible, it is possible to find different site wiring to that shown.
d) Relay diagram is inside cover. Relay can be replaced by double pole changeover relay (DPDT).
e) Note that neutral on the actuator is black.
f) Connect frost thermostat across junction box terminals L–6.

Figure 11.41 *Drayton Flowshare 5 system with TA/M4 actuator, RB2 relay box and programmer*

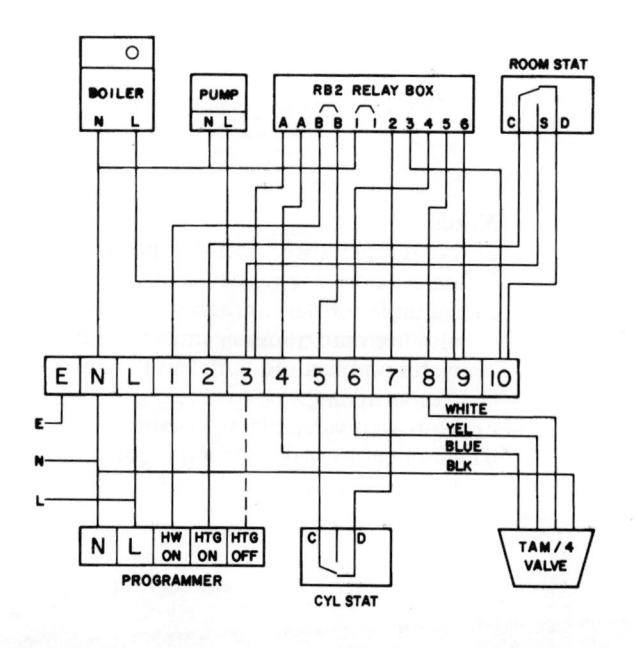

Notes

a) Suitable for Basic programmers only unless programmer has 'Heating Off' facility and dotted line is connected.
b) If a time switch is used link 'Heating On' and 'Hot Water On'.
c) As this actuator is reversible, it is possible to find different site wiring to that shown.
d) Relay diagram is inside cover. Relay can be replaced by double pole changeover relay (DPDT).
e) Note that neutral on the actuator is black.
f) B–B and 1–1 are internally linked but A–A are not and must remain unlinked.
g) Connect frost thermostat across junction box terminals 1–6.

204

Wiring system diagrams

Figure 11.42 *Drayton Plan 7 system with two TA/M2A actuators providing full temperature and programming control*

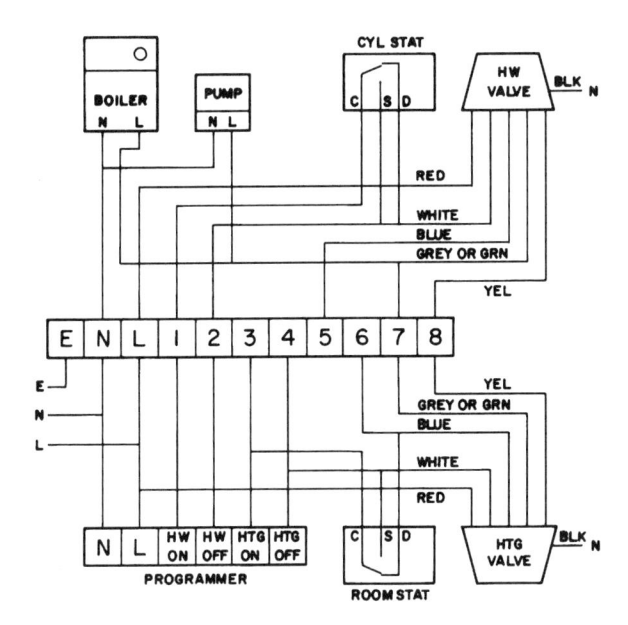

Notes

a) Essential that programmer has 'Off' signals to both heating and hot water.
b) System requires changeover room and cylinder thermostats.
c) Note that black is neutral and green is a live conductor.
d) Connect frost thermostat across junction box terminals L–6.

Figure 11.43 *Homewarm Manual System*

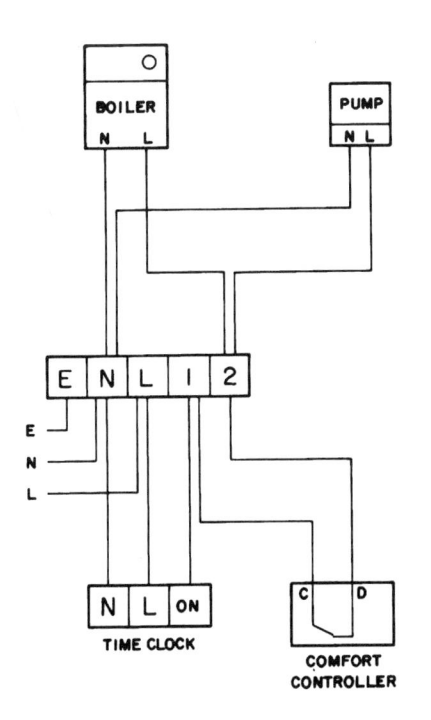

Notes

a) A manually operated 3-port valve is used to divert flow to heating or hot water circuits as required.
b) If no comfort controller, link 1–2 in junction box.
c) Connect frost thermostat across junction box terminals L–2.

Installing and Servicing Domestic Heating Wiring Systems and Control

Figure 11.44   *Homewarm Auto System with Switchmaster VM5 actuator*

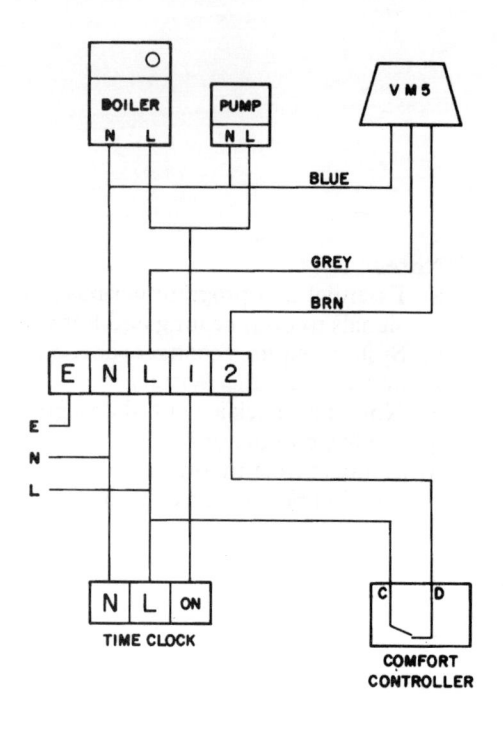

Notes

a) Designed as a low-cost installation. Refer to motorized valve section (Chapter 6, Switchmaster VM5).
b) If an orange wire is exposed in valve flex, this can be cut off and disregarded.
c) Connect frost thermostat across junction box terminals L–1.

Figure 11.45   *Honeywell Plan System using the V4073 6-wire, 3-port motorized valve with integral relay giving full temperature control*

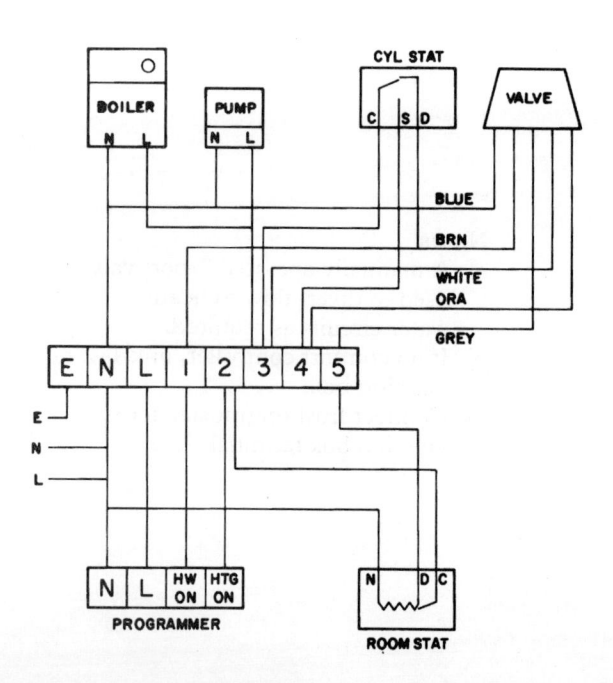

Notes

a) Suitable in this diagram for a Basic programmer only.
b) Connect frost thermostat across junction box terminals L–3.

Wiring system diagrams

Figure 11.46 *Honeywell Y Plan System using the V4073 6-wire, 3-port motorized valve with integral relay giving full temperature and programming control*

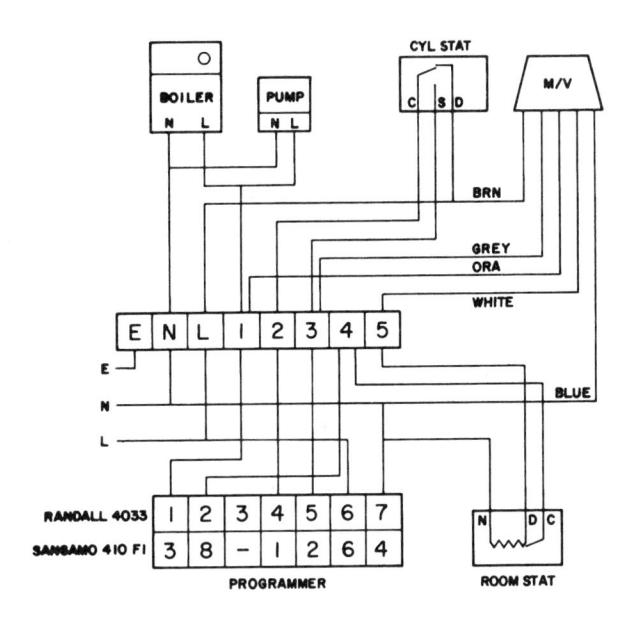

Notes

a) Suitable in this diagram for Full control using programmers shown.
b) For more information on the V4073 6-wire motorized valve refer to Chapter 6.
c) Connect frost thermostat across junction box terminals L–3.

Figure 11.47 *Landis & Gyr LGM System*

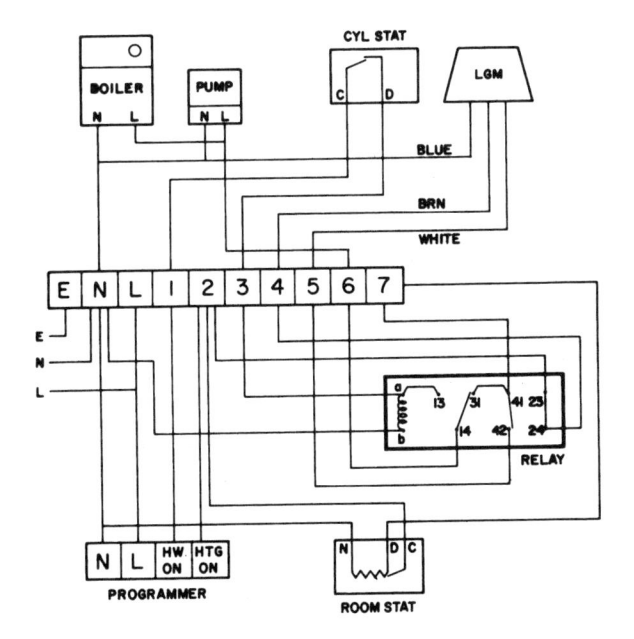

Notes

a) Suitable for Basic or Full control, depending on programmer used.
b) Connect frost thermostat across junction box terminals L–7.

207

Installing and Servicing Domestic Heating Wiring Systems and Control

Figure 11.48 *SMC Control Pack 2 system with one boiler and a pump to both central heating and hot water circuits. No motorized valves. Provides full temperature and programming control*

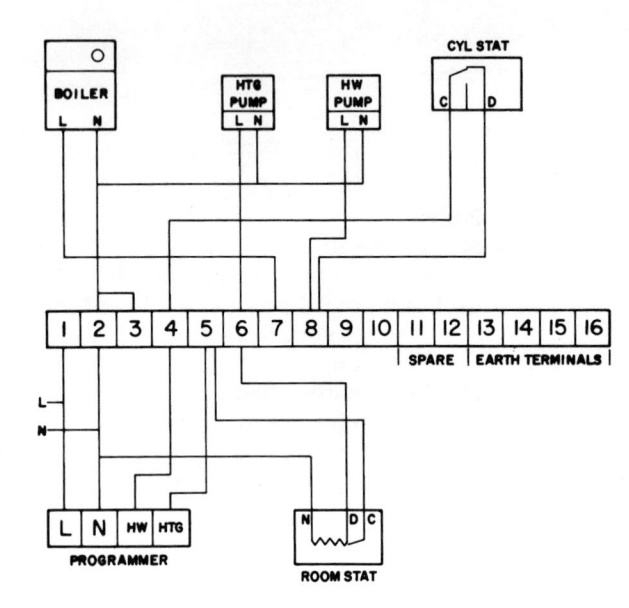

Notes
a) All controls wired into SMC wiring centre incorporating relay.
b) Internal wiring of SMC wiring centre not shown.
c) Suitable for Basic or Full programming.
d) Terminals 2–3 are linked internally for neutral connection.
e) Connect frost thermostat across junction box terminals L–6.

Figure 11.49 *SMC Control Pack 2 system with pump overrun boiler and pump to both central heating and hot water circuits. No motorized valves. Provides full temperature and programming control*

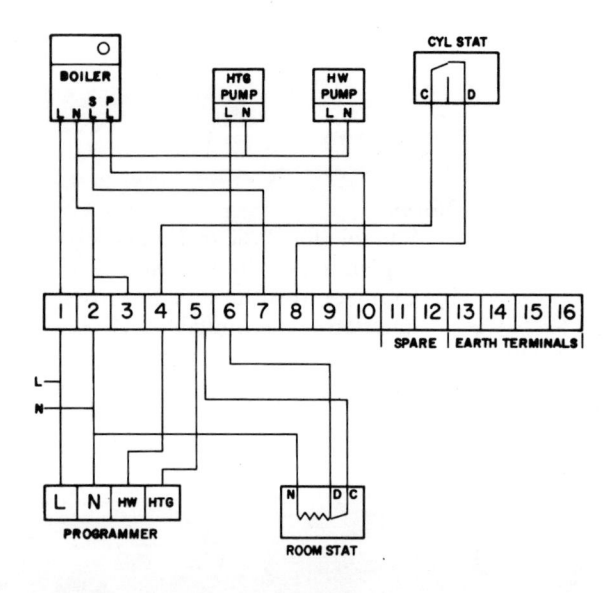

Notes
a) All controls wired into SMC wiring centre incorporating relay.
b) Internal wiring of SMC wiring centre not shown.
c) Suitable for Basic or Full programming.
d) Terminals 2–3 are linked internally for neutral connection.
e) Connect frost thermostat across junction box terminals L–6.

Wiring system diagrams

Figure 11.50 *Sunvic Duoflow System with RJ 1801 relay box and programmer*

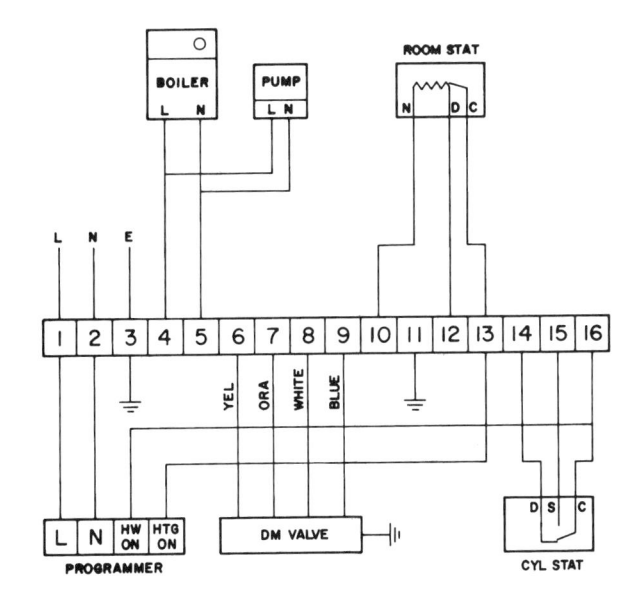

Notes

a) Suitable for Basic programming only.
b) Yellow CH/HW, orange HW, white CH, blue N.
c) Can be used with any DM actuator.
d) Where a programmer is used, remove links 1–13 and 13–16.
e) Where a time switch is used, remove links 13–16.
f) Where there is no time control leave links in place.
g) Connect frost thermostat across junction box terminals 1–12.

Figure 11.51 *Sunvic Duoflow System with RJ 2801 relay box and programmer*

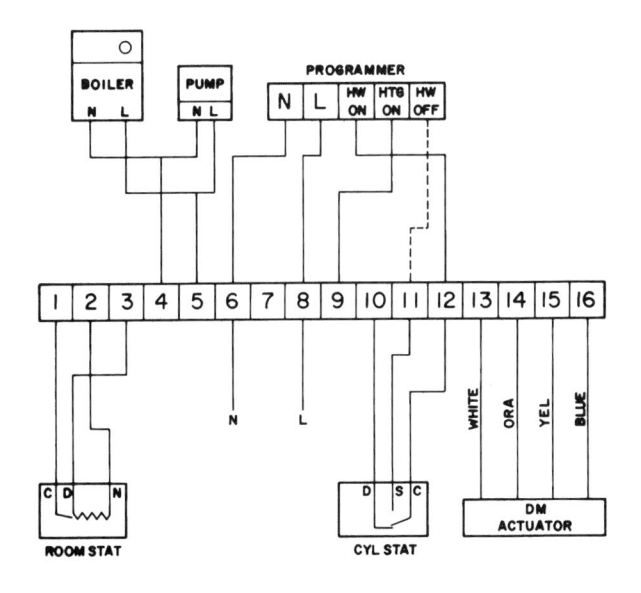

Notes

a) Suitable for Basic programmer only unless programmer has 'Central Heating Only' and 'Heating Off' signal, in which case dotted line must be connected.
b) Can be used with any DM actuator.
c) Connect frost thermostat across junction box terminals 3–8.

Installing and Servicing Domestic Heating Wiring Systems and Control

Figure 11.52 Sunvic Duoflow System with RJ 2802 or RJ 2852 relay box and time clock

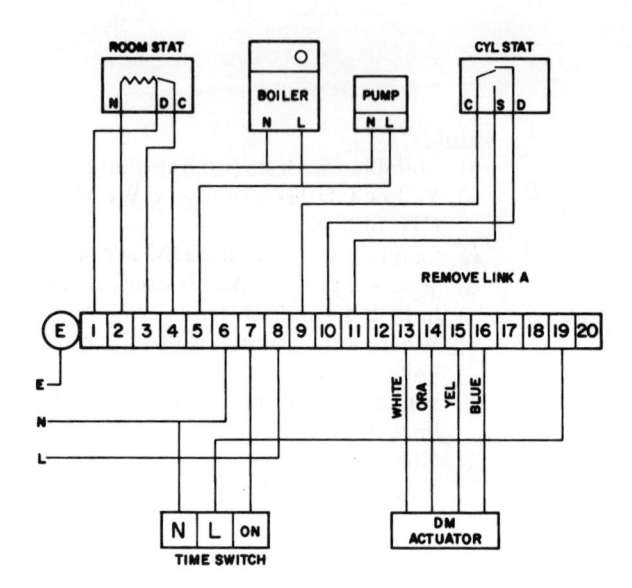

Notes
a) Remove Link A.
b) Can be used with any of the DM actuators.
c) RJ 2852 is the plug-in version of the RJ 2802.
d) Connect frost thermostat across junction box terminals 1–19.

Figure 11.53 Sunvic Duoflow System with RJ 2802 or RJ 2852 relay box and programmer

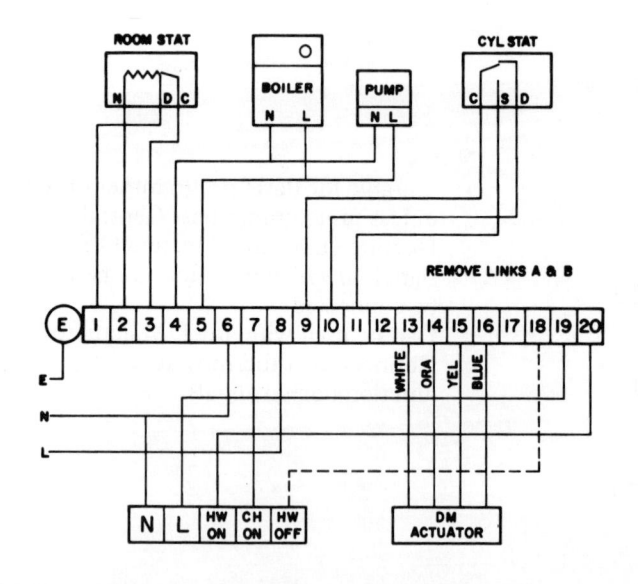

Notes
a) Remove links A and B.
b) Suitable for Basic programming only unless programmer has 'Central Heating Only' and 'Heating Off' signal, in which case dotted line must be connected.
c) Can be used with any of the DM actuators.
d) RJ 2852 is the plug-in version of the RJ 2802.
e) Connect frost thermostat across junction box terminals 1–19.

Wiring system diagrams

Figure 11.54 *Switchmaster Midi System*

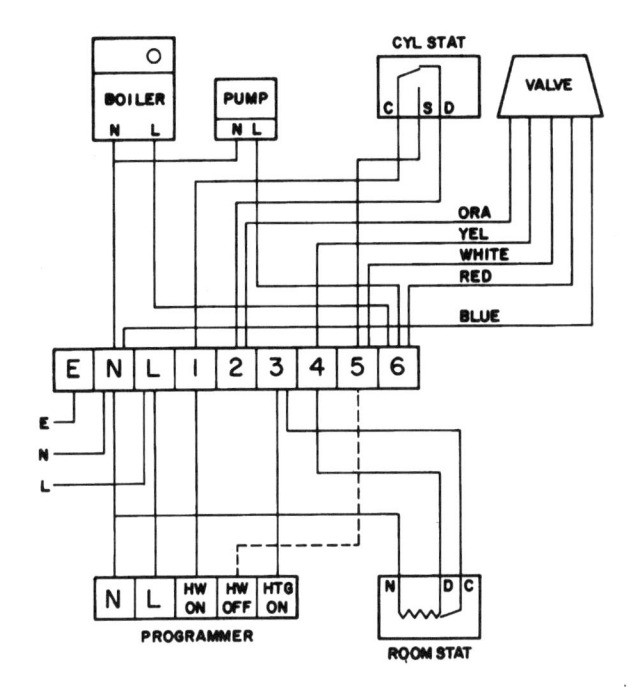

Notes
a) Suitable for Basic control only unless programmer has 'Hot Water Off' signal, in which case full control is possible by connecting dotted line.
b) Connect frost thermostat across junction box terminals L–4.

Various frost protection thermostat wiring

Figure 11.55 *Wiring of a double pole frost thermostat to a gravity hot water, pumped central heating system*

Figure 11.56 *Wiring of a single pole frost thermostat to a gravity hot water, pumped central heating system using a DPDT relay*

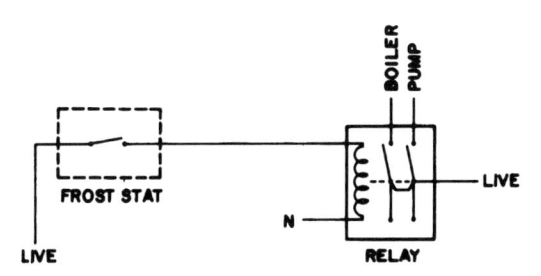

211

Installing and Servicing Domestic Heating Wiring Systems and Control

Figure 11.57 *Wiring a frost thermostat to a motor open, motor closed motorized valve*

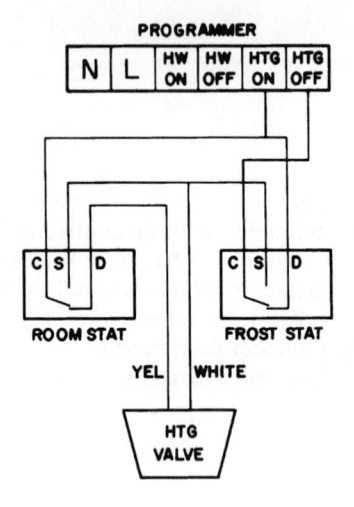

Notes

a) Will only function if room thermostat is also calling for heat.
b) To connect into existing, remove wire from programmer 'Heating Off' to room thermostat 'Satisfied'. Wire from programmer 'Heating Off' to frost thermostat 'Common'. Wire frost thermostat 'Satisfied' to room thermostat 'Satisfied' and wire frost thermostat 'Demand' to room thermostat 'Common'.
c) Requires SPDT room and frost thermostats.

Figure 11.58 *Wiring a frost thermostat to a fully pumped system using two motor open, motor close motorized valves*

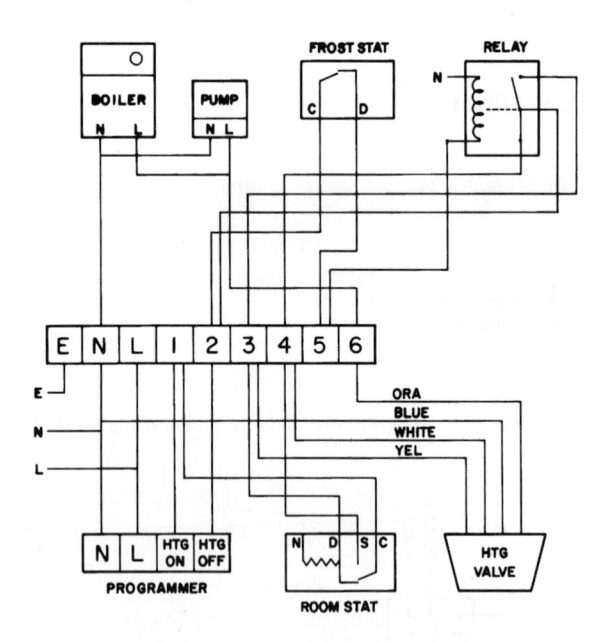

Notes

a) Frost thermostat shown on heating circuit but can be used on hot water if required.
b) Requires an SPDT replay.
c) Requires an SPST frost thermostat.
d) Frost thermostat common can go to terminal L if required instead of 'Heating Off'.

Supplementary wiring diagrams

Figure 11.59 *Utilizing a full control programmer for basic control*

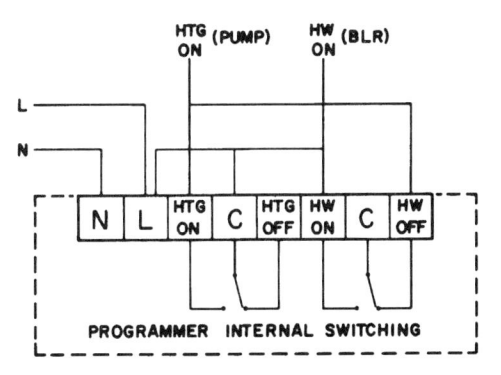

Note
Programmer must have voltage free terminals

Figure 11.60 *Addition of a changeover switch to a priority system incorporating a 3-port valve, e.g. Honeywell V4044, to enable changing from hot water or heating priority as required. A simple 2-way light switch is ideal for this purpose.*

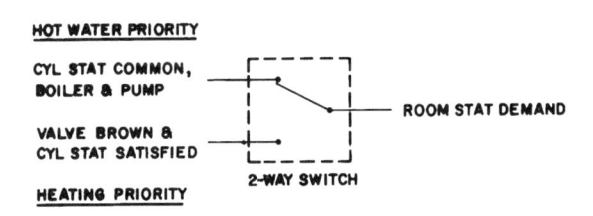

Figure 11.61 *Simple pump overrun thermostat wiring*

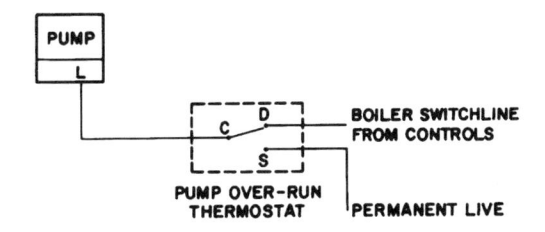

Note
Could be used as an additional control, e.g. on cast iron boilers, by employing a pipe thermostat located on the flow or return pipework as required.

Relays

Relays are often viewed with suspicion and a lot of electricians will hold their hands up in horror at the sight of a relay, especially when included into what already looks a complicated heating system. Relays deployed in this way are usually for switching 240V to get over problems of 'back-feed', which may cause the system to do strange things. They can also be used for switching different voltages, e.g. a 240V coil could switch 24V and vice versa. There are a number of diagrams in the book which require the use of a relay and it may well be that the fault the engineer has been sent to find exists because a relay has never been fitted.

All relays are shown relaxed or de-energized.

ABBREVIATIONS:
SPDT   Single pole double throw – top
DPDT   Double pole double throw – middle
COM   Common
NO   Normally open – de-energized
NC   Normally closed – de-energized

Figure 11.62 *Wiring of relays, including examples*

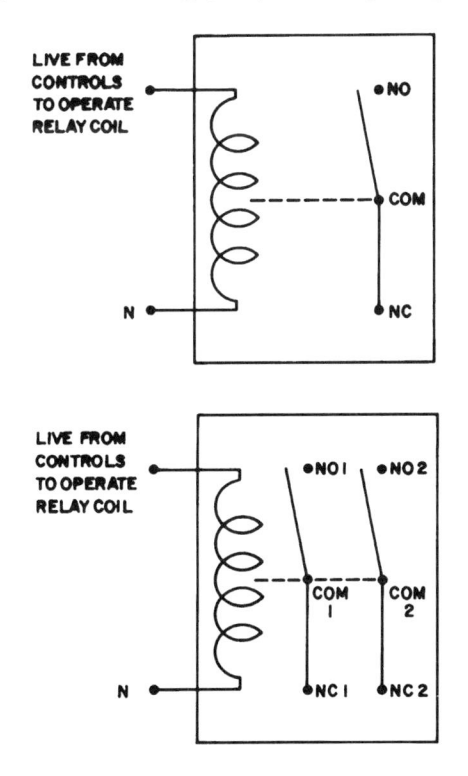

Installing and Servicing Domestic Heating Wiring Systems and Control

Figure 11.63   *Wiring of a relay to allow a 2-wire (SPST) cylinder or room thermostat to function as a 3-wire (SPDT)*

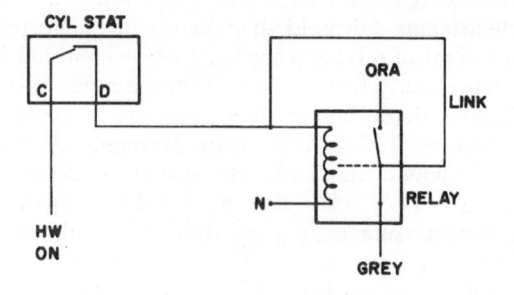

Notes

a) Ideal for when two wires have been run to a room or cylinder thermostat and system alterations require three wires.
b) Shown on a 3-way valve with standard flex conductors.

Simplified warm air unit wiring

Figure 11.64

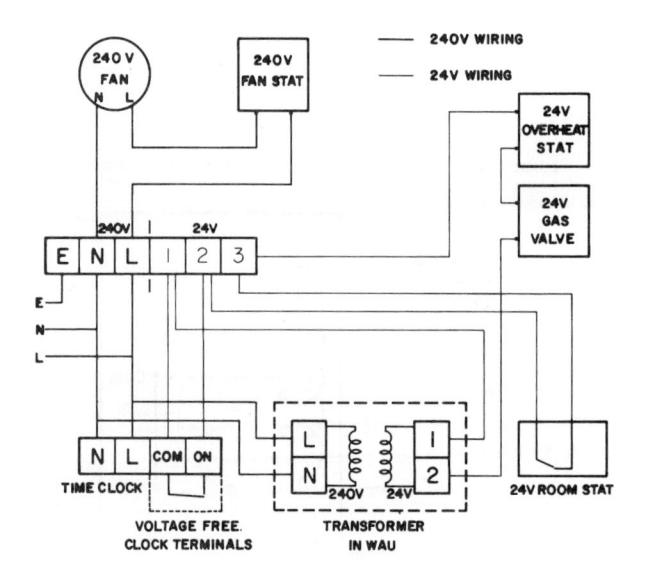

Notes

When the clock and room thermostat contacts are made, the main gas burner will light, warming the heat exchanger. The heat will be detected by the fan thermostat which will turn the fan on.

When the clock or room thermostat contacts break, the main gas valve will go off and the fan will continue to run for a while to clear the residual heat from the heat exchanger, until it falls below the setting of the fan thermostat.

It is usually essential that the time clock has voltage free terminals and in some clocks it is necessary to remove a link to achieve this.

12

Interchangeability guide for programmers and time switches

This guide is to assist the replacement of faulty or obsolete programmers and time switches but it must be remembered that each group relates only to backplate and wiring, or product, and not to the facilities provided by the programmer or time switch itself, although a brief indication of setting options is given. Full details of the various differences, including dimensions, can be found earlier in this book.

This guide will also help in providing the householder with a prompt replacement, and in some cases offer the opportunity for replacing an electromechanical programmer with an electronic equivalent providing more programming options with minimum inconvenience. Obvious groupings are not included. For example there are four different models of Randall 103 time clock but are all wired the same and are interchangeable. Any differences are indicated earlier in this book.

The following groups of programmers and time switches can usually be directly interchanged without changing the backplate. However, backplate design may vary making a straight exchange impossible, e.g. a Landis & Gyr RWB2 will not fit an ACL 722 backplate but the reverse is possible. This is due to the provision of an earth terminal on the ACL backplate. Note also that it would be good working practice to change the backplate, as occasionally the contacts become worn and give rise to a fault situation that could mislead the engineer into suspecting that the programmer or time switch is faulty. In some cases it may be necessary to make a minor wiring alteration, and this is included in this chapter.

Programmers

Please note that all programmers listed below are suitable for use on Basic and Full systems unless stated otherwise. Therefore when changing ensure that the new programmer is set for the same system as the old one.

Group A
Electronic

ACL LP 112 – 24 hour programming
ACL LP 241 – 24 hour programming
ACL LP 522 – 5/2 day programming
ACL LP 722 – 7 day programming
ACL LS 112★ – Basic systems only, 24 hour programming
ACL LS 241 – 24 hour programming
ACL LS 522 – 5/2 day programming
ACL LS 722 – 7 day programming
Barlo EPR 1 – 5/2 day programming
Danfoss Randall CP15 – 24 hour or 5/2 day programming

Danfoss Randall CP75 – 5/2 day or 7 day programming
Danfoss Randall FP15 – 24 hour or 5/2 day programming
Danfoss Randall FP75 – 5/2 day or 7 day programming
Danfoss Randall MP15 – Basic systems only
   – 24 hour or 5/2 day programming
Danfoss Randall MP75 – Basic systems only
   – 5/2 day or 7 day programming
Drayton Tempus 3 – 24 hour programming
Drayton Tempus 4 – 5/2 day programming
Drayton Tempus 7 – 7 day programming
Honeywell ST 6200 – Basic systems only, 24 hour programming
Honeywell ST 6300 – Full systems only, 24 hour programming
Honeywell ST 6400 – Full systems only, 7 day programming
Honeywell ST 6450 – Full systems only, 5/2 day programming
Landis & Gyr RWB 20 – 7 day programming
Landis & Gyr RWB 40 – 24 hour programming
Landis & Gyr RWB 102 – Basic systems only, 24 hour programming
Landis & Gyr RW 200 – 24 hour programming
Landis & Gyr RWB 252 – 5/2 day programming
Landis & Steafa RWB 9 – 24 hour, 5/2 day, 7 day programming
Potterton Mini-minder E – 24 hour programming
Smiths Controller 1000 – 24 hour programming
Sunvic 207 – 24 hour, 5/2 day, 7 day programming
Tower DP 72 – 7 day programming
Tower QE2 – 7 day programming
*A wiring alteration will be necessary.

Electromechanical

Crossling Controller
Glow-Worm Mastermind
Landis & Gyr RWB 1
Landis & Gyr RWB 2 MK 1
Landis & Gyr RWB 2 MK 2
Landis & Gyr RWB 2.9 – no neon indicators
Potterton Mini-minder

Group B

Electronic

Danfoss Randall FP 965 – 5/2 day or 7 day programming
Horstmann 525 – 24 hour programming
Horstmann 527 – 7 day programming
Horstmann H21 – 24 hour programming
Horstmann H121 – 24 hour programming
Horstmann H27 – 7 day programming
Randall 922 – 24 hour programming
Randall 972 – 7 day programming
Randall Set 2★★ – Basic systems only, 24 hour programming
Randall Set 2E★★ – Basic systems only, 24 hour programming

Randall Set 3 – Full systems only, 24 hour programming
Randall Set 3E – 24 hour programming
Randall Set 5 – 5/2 day programming
**Link 1–5 on backplate.

Electromechanical

Danfoss 3002
Horstmann 425 Diadem
Horstmann 425 Tiara – no neon
Randall set 3M

Note: Only the Danfos Randall FP 965 will fit the Randall 922 and 972 backplate.

Group C

Electronic

Honeywell ST 499A – Full systems only, 24 hour programming
Honeywell ST 699B – Full systems only, 24 hour programming
Honeywell ST 699C – Full systems only, 24 hour programming
Thorn Microtimer – Full systems only, 24 hour programming

Group D

Electronic

Potterton EP 2000 – 24 hour programming
Potterton EP 2001 – 5/2 day programming
Potterton EP 2002 – 5/2 day programming
Potterton EP 3000 – 7 day programming
Potterton EP 3001 – 7 day programming
Potterton EP 3002 – 7 day programming
Potterton EP 6000 – 7 day and 5/2 day programming (optional each channel)
Potterton EP 6002 – full systems only, 7 day programming

Group E

Electronic

Switchmaster 9000 – 24 hour programming
Switchmaster 9001 – 24 hour programming

Electromechanical

Switchmaster 900
Switchmaster 905

Time switches

Group A

Electronic

Horstmann H11 – 24 hour programming
Horstmann H17 – 7 day programming

Electromechanical

Danfoss 3001
Horstmann 425 Coronet

Group B

Electronic

Danfoss Randall TS975 – 5/2 day or 7 day programming
Randall 911 – 24 hour programming
Randall 971 – 7 day programming
Randall Set 1 – 1E – 24 hour programming
Randall Set 4 – 5/2 day programming
Sangamo Set 1 – 24 hour programming

Group C

See note after Group G.

Electronic

Landis & Gyr RWB 50 – 24 hour programming
Landis & Gyr RWB 100 – 24 hour programming
Landis & Gyr RWB 152 – 5/2 day programming
Landis & Staefa RWB 7 – 24 hour, 5/2 day, 7 day programming
Potterton Mini-minder ES – 24 hour programming

Electromechanical

Landis & Gyr RWB 30

Group D

See note after Group G.

Electronic

ACL LP 111 – 24 hour programming
ACL LP 711 – 7 day programming

ACL LS 111 – 24 hour programming
ACL LS 711 – 7 day programming
Sunvic 107 – 24 hour, 5/2 day, 7 day programming

Group E

See note after Group G.

Electronic

Drayton Tempus 1 – 24 hour programming
Drayton Tempus 2 – 5/2 day programming

Group F

See note after Group G.

Electronic

Tower DT 71 – 7 day programming
Tower QE 1 – 7 day programming

Group G

Electronic

Danfoss Randall TS 15 – 24 hour, 5/2 day programming
Danfoss Randall TS 75 – 5/2 day, 7 day programming
Honeywell ST 6100A – 24 hour programming
Honeywell ST 6100C – 7 day programming

Note: All time switches in Groups C, D, E, F and G have similar backplates but a wiring alteration will also be required.

Group H

Electronic

Potterton EP 4000 – 7 day programming
Potterton EP 4001 – 5/2 day programming
Potterton EP 4002 – 5/2 day programming
Potterton EP 5001 – 7 day programming
Potterton EP 5002 – 7 day programming

13

Manufacturers' trade names and directory

ACL Drayton	See Appliance Components Ltd
Aga	All enquiries to Glywed Consumer and Building Products
Agaheat	P.O. Box 30, Ketley, Telford, Shropshire, TF1 4DD
Aga-Rayburn	Tel: 01952 641100 Fax: 01952 641961
Alde Int. (UK) Ltd.	Sandfield Close, Moulton Park, Northampton, NN3 1AB
	Tel: 01604 494193 Fax: 01604 499551
Allied Heatwave	All enquires to Glywed Consumer and Building Products
Allied Ironfounders Ltd	All enquiries to Glywed Consumer and Building Products
Alpha-Ocean Boilers	See Alphatherm Ltd
Alphatherm Ltd	United House, Goldsel Road, Swanley, Kent, BR8 8EX
	Tel: 01322 665522 (helpline) 01322 669443 Fax: 01322 615017
Altenic Ltd	Airfield Industrial Estate, Hixon, Staffs, ST18 0PF
	Tel: 01889 271371 Fax: 01889 270577
AMF-Venner	All enquiries to Heating Control Services
	Tel: 01922 34503
APECS Ltd	Blackburn Road, Clayton-Le-Moors, Accrington, BB5 5JW
	Tel: 01254 872707 Fax: 01254 392055
Appliance Components Ltd	Cordwallis Street, Maidenhead, Berks SL6 7BQ
	Tel: 01895 444012 Fax: 01628 75062
Ariston Boilers	All enquiries to MTS (GB) Ltd
Atlantic Boilers	P.O. Box 11, Ashton Under Lyne, Lancs, OL6 7TR
	Tel: 0161 344 5664 (technical) 0161 330 4422 Fax: 0161 339 8136
AWB	All enquiries to Time and Temperature
Barlo Products Ltd.	Barlo House, Finway, Dallow Road, Luton, Beds, LU1 1TR
	Tel: 01582 480333 (technical) 01582 482580 Fax: 01582 402459
Baxi Heating	Brownedge Road, Bamber Bridge, Preston, Lancs, PR5 6SN
	Tel: 01772 336201 Fax: 01772 315998
BBC Industries Ltd	21 Northgate Street, Devizes, Wilts, SN10 1JL
	Tel: 01380 727333 Fax: 01380 727666
Boulter Boilers Ltd	Magnet Works, White House Road, Ipswich, IP1 5JA
	Tel: 01473 241555 Fax: 01473 241321
Brassware Ferroli	All enquiries to Ferroli
Burco Dean Domestic Appliances	Rose Grove, Burnley, Lancs, BB12 6AL
	Tel: 01282 427241 Fax: 01282 831206
Burco-Maxol	See Burco-Dean Domestic Appliances
Caradon-Ideal Heating Ltd	P.O. Box 103, National Avenue, Hull, HU5 4JN
	Tel: 01482 492251 (helpline) 01482 498603 Fax: 01482 448858
R & S Cartwright (Manchester) Engineering Ltd	Floats Road, Roundthorn Industrial Estate, Manchester, M23 9NE
	Tel: 0161 998 9829 Fax: 0161 946 0738
Centurion Gas Products Ltd	P.O.Box 323, Ipswich, Suffolk, IP1 5JB
	Tel: 01473 747472 Fax: 01473 747479
Chaffoteaux et Maury	Trench Lock, Telford, Shropshire, TF1 4SZ
	Tel: 01952 222727 Fax: 01952 243493

Manufacturers' trade names and directory

Church Hill Systems Ltd	Frolesworth, Lutterworth, Leics, LE17 5EC Tel: 01455 202314 Fax: 01455 202606
G. R. Claudio Ltd	Clarisham House, Morson Road, Enfield, Middx, EN3 4NQ Tel: 0181 804 7202 Fax: 0181 804 8163
Clyde Combustions Ltd	Cox Lane, Chessington, Surrey, KT9 1SL Tel: 0181 391 2020 Fax: 0181 397 4598
Clyde Valley Control Systems Ltd	33 Glenburn Rd, College Milton North, East Kilbride, G74 3BA Tel: 01355 247921 Fax: 01355 249197
Combat Engineering Ltd	Oxford Street, Bilston, West Midlands, WV14 7EG Tel: 0192 494425 Fax: 01902 403200
Crosslee PLC	Lightcliffe Fact, Hipperholme, Halifax, West Yorks, HX3 8DE Tel: 01422 203963 Fax: 01422 204475
Danfoss	All enquiries to Danfoss-Randall
Danfoss-Randall	Ampthill Road, Bedford, Beds, MK42 9ER Tel: 01234 364621 Fax: 01234 219705
Drayton Controls	Now ACL-Drayton Chantry Close, West Drayton, Middx, UB7 7SP Tel: 01895 444012 Fax: 01895 421901
Dunphy Oil & Gas Burners Ltd	Queensway, Rochdale, Lancs, OL11 2SL Tel: 01706 649217 Fax: 01706 55512
Eastham Maxol	See Burco Dean Domestic Appliances
Eberle Controls	25 Gosforth Close, Sandy Business Park, Sandy, Beds, SG19 1RB Tel: 01767 692323 Fax: 01767 692333
ELM Leblanc Ltd	12 Chesterfield Way, Hayes, Middx, UB3 3NW Tel: 0181 848 7522 Fax: 0181 848 1984
Elson	See Elsy & Gibons Ltd
Elsy & Gibbons Ltd	Simonside, South Shields, Tyne & Wear, NE34 9PE Tel: 0191 427 0777 Fax: 0191 427 0888
Esse (1984) Ltd	All enquiries to Ouzledale Foundry Co. Ltd
Euramo	See Wilo Salmson Pumps Ltd
Eurocombi	See MTS (GB) Ltd
Eurotronics	Unit 16, Monument Industrial Park, Chalgrove, Oxon, OX44 7RW Tel: 01865 400526 Fax: 01865 400524
Ferroli	Stockton Close, Minworth Industrial Park, Minworth, Sutton Coldfield, West Midlands, B76 8DH Tel: 0121 352 3500 Fax: 0121 353 3505
Firefly (UK) Ltd	Unit 4, Stag Business Park, Christchurch Road, Ringwood, Hants, BH24 3SB Tel: 01425 480210 Fa: 01425 479089
Flash	See Eurotronics
GEC Nightstor	See APECS Ltd
Gemini Heat Systems Ltd	All enquiries to Grant Engineering (UK) Ld
Gledhill Water Storage Ltd	Sycamore Estate, Squires Gate, Blackpool, Lancs, FY4 3RL Tel: 01253 401494 Fax: 01253 474445
Glow Worm	See Hepworth Heating Ltd
Glynwed Consumer & Building Products	P.O. Box 30, Ketley, Telford, Shropshire, TF1 4DD Tel: 01952 642000 Fax; 01952 641961
G P Burners Ltd	2D Hargreaves Road, Groundwell Industrial Estate, Swindon, Wilts, SN2 5AZ Tel: 01793 705085 Fax: 01793 705263
Grant Engineering (UK) Ltd	Brunel Road, Churchfields, Salisbury, Wilts, SP2 7PU Tel: 01722 412284 Fax: 01722 323003

Grasslin (UK) Ltd	Vale Rise, Tonbridge, Kent, TN9 1TB
	Tel: 01732 359888 Fax: 01732 354445
Grundfos Pumps Ltd	Grovebury Road, Leighton Buzzard, Beds, LU 8TL
	Tel: 01525 850000 Fax: 01525 850011
Halstead Boilers Ltd	4 First Avenue, Bluebridge Industrial Estate, Halstead, Essex, CO9 2EX
	Tel: 01787 475557 Fax: 01787 474588
Hamworthy Engineering Ltd	Fleets Corner, Poole, Dorset, BH17 7LA
	Tel: 01202 665566 Fax: 01202 665111
Hart-Lonex	All enquiries to Lonex Ltd
Hattersley Brothers	All enquiries to Caradon-Ideal Ltd
Heating World Group Ltd	Excelsior Works, Eyre Street, Birmingham, B18 7AD
	Tel: 0121 454 2244 Fax: 0121 454 4488
Heatrae-Sadia Ltd	Hurricane Way, Norwich, Norfolk, NR6 6EA
	Tel: 01603 424144 Fax: 01603 409409
Heb Boilers Ltd	See Thermecon
Hepworth Heating Ltd	Nottingham Road, Belper, Derbyshire, DE6 1JT
	Tel: 01773 824141 (technical services) 01773 828100 Fax: 01773 820569
Honeywell Control Systems Ltd	Residential Division, Charles Square, Bracknell, Berks, RG12 1EB
	Tel: 01344 424555 (technical support) 0345 678999
	Fax: 01344 416416
Horstmann Timers & Controls Ltd	Newbridge Road, Bath, BA1 3EF
	Tel: 01225 421141 Fax: 01225 423070
Hoval Ltd	Northgate, Newark, Notts, NG24 1JN
	Tel: 01636 72711 Fax: 01636 73532
HRM Boiler Co	Haverscroft Industrial Estate, Attleborough, Norfolk, NR17 1YE
	Tel: 01953 455400 Fax: 01953 454483
IMI Pactrol Ltd	10 Pithey Place, West Pimbo, Skelmersdale, Lancs, WN8 9PS
	Tel: 01695 725152 Fax: 01695 724400
IMI Range	See IMI Waterheating
IMI Waterheating	P.O. Box 1, Bridge Street, Stalybridge, Cheshire, SK15 1PQ
	Tel: 0161 338 3353 Fax: 0161 303 2634
Imstor	See Church Hill Systems Ltd
International Janitor	All enquiries to Potterton Myson (Gateshead Office)
JLB Group	All enquiries to Crosslee PLC
Johnson & Starley Ltd	Rhosili Road, Brackmills, Northampton, NN4 0LZ
	Tel: 01604 762881 Fax: 01604 767408
Keston Boilers	34 West Common Road, Hayes, Bromley, Kent, BR2 7BX
	Tel: 0181 462 0262 Fax: 0181 462 4459
Landis & Gyr	All enquiries to Landis & Staefa
Landis & Staefa (UK) Ltd	Hortonwood 30, Telford, Shropshire, TF1 4ET
	Tel: 01952 602048 Fax: 01952 602059
Leblanc	See ELM Leblanc
Lennox Industries Ltd	P.O. Box 174, Westgate Interchange, Northampton, NN5 5AG
	Tel: 01604 591159 Fax: 01604 587536
Lonex Ltd	Baynes Mews, London, NW3 5BX
	Tel: 0171 794 9554 Fax: 0171 431 2261
Malvern Boilers Ltd	Spring Lane North, Malvern, Worcs, WR14 1BW
	Tel: 01684 893777 Fax: 01684 893776
Maxol	All enquiries to Burco Dean Domestic Appliances
Maxton	All enquiries to Potterton-Myson (Gateshead Office)
Meta	See Modular Heating Sales Ltd

Manufacturers' trade names and directory

Modular Heating Sales Ltd	35 Nobel Square, Burnt Mills Industrial Estate, Basildon, Essex, SS13 1LT Tel: 01268 591010 Fax: 01268 728202
MTS (GB) Ltd	MTS Building, Hughenden Ave, High Wycombe, Bucks, HP13 5FT Tel: 01494 459711 Fax: 01494 459775
Myson Combustion Products Ltd	All enquiries to Potterton Myson (Gateshead Office)
Myson Heating	All enquiries to Potterton Myson (Gateshead Office)
Nailmere Ltd (AWB & Pyrocraft Boilers)	All enquiries to Time & Temperature
Nu-Way Ltd	P.O. Box 1, Vines Lane, Droitwich, Worcs, WR9 8NA Tel: 01905 794331 Fax: 01905 794017
Ocean Boilers	See Alphatherm Ltd
Offergram	Ceased trading – spares available from usual sources
Ouzledale Foundry Co. Ltd	Long Ing, Barnoldswick, Colne, Lancs, BB8 6BN Tel: 01282 813235 Fax: 01282 816876
Pakaway Perrymatics	See Offergram
Paragon Electric Ltd	All enquiries to Heating Control Services Tel: 01922 34503
Parkray	All enquiries to Hepworth Heating Ltd
Pegler Ltd	St Catherines Avenue, Doncaster, DN4 8DF Tel: 01302 368581 Fax: 01302 367661
Perrymatic	See Offergram
Potterton Myson Ltd	Eastern Avenue, Team Valley Trading Estate, Gateshead, Tyne & Wear, NE11 0PG Tel: 0191 491 4466 Fax: 0191 491 7568
Potterton Myson Ltd	Portobello Works, Emscote Road, Warwick, CV34 5QV Tel: (service) 01926 496896 (technical helpline) 01926 410044 Fax: 01926 410006
Powell Duffryn	All enquiries to Potterton Myson (Gateshead Office)
Powermax	See IMI Waterheating
Porermatic Ltd	Winterhay Lane, Ilminster, Somerset, TA19 9PQ Tel: 01460 53535 Fax: 01460 52341
Prism Energy Technology Ltd	1 Hanham Business Park, Memorial Road, Hanham, Bristol, BS15 3JE Tel: 0117 949 8800 Fax: 0117 949 8888
Proscon	All enquiries to Pullin Electronics Ltd
Pullin Electronics Ltd	All written enquiries to Heatrae-Sadia Tel: 01603 404848
Pyrocraft	All enquiries to Time & Temperature
Radiant Boilers (Agent)	Unit 16, Spurlings Yard, Wallington, Gareham, Hants, PO17 6AB Tel: 01329 828 555
Radiation	All enquiries to Hepworth Heating Ltd
Randall Electronics	All enquiries to Danfoss-Randall
Raenheat	Chartists Way, Morley, Leeds, LS27 9ET Tel: 0113 252 7007 Fax: 0113 259 7713
Rayburn	All enquiries to Glynwed Consumer and Building Products
Redring Electric Ltd	Celta Road, Peterborough, PE2 9JJ Tel: 01733 313213 (technical 01733 314057) Fax: 01733 310606
Riello Ltd	The Ermine Centre, Ermine Business Park, Huntingdon, Cambs, PE18 6XX Tel: 01480 432144 Fax: 01480 432191

Ringdale UK Ltd	56 Victoria Road, Burgess Hill, West Sussex, RH15 9LR
	Tel: 01444 871349 Fax: 01444 870228
Saacke Ltd	Marshlands Spur, Farlington, Portsmouth, Hants, PO6 1RX
	Tel: 01705 383111 Fax: 01705 327120
Sangamo	See Schlumberger
Satchwell Control Systems Ltd	P.O. Box 57, Farnham Road, Slough, Berks, SL1 4UH
	Tel: 01753 550550 Fax: 01753 824078
Satchwell Sunvic (domestic)	See Pegler or Sunvic
Sauter Automation Ltd	Inova House, Hampshire International Business Park, Crockford Lane, Chineham, Basingstone, RG24 8WH
	Tel: 01256 374400 Fax: 01256 374455
Schlumberger	Industrial Estate, Port Glasgow, Renfrewshire, PA14 5XG
	Tel: 01475 745131 Fax: 01475 744567
Seagoe Technology	221 Europe Boulevard, Gemini Business Park, Warrington, Cheshire, WA5 5TN
	Tel: 01762 333131 Fax: 01925 232624
Sime Heating Products UK Ltd	Jubilee Works, Middlecroft Road, Staveley, Chesterfield, Derbyshire, S43 3XN
	Tel: 01246 471950 Fax: 01246 281822
Sinclair	See Halstead Heating
Smith & Wellstood	All enquiries to Ouzledale Foundry Co. Ltd
Smith Meters	See Appliance Components Ltd
Smiths Industries	All enquiries to Timeguard
Solaire Ltd	All enquiries to Ordermark, 67 Charminster Road, Bournemouth, Dorset, BH8 8UE
	Tel: 01202 551666 Fax: 01202 295666
Sopac-Jaeger Controls Ltd	17 Invincible Road, Farnborough, Hants, GU14 7QN
	Tel: 01252 511981 Fax: 01252 524018
Southern Digital Ltd	No longer trading
Stelrad-Ideal	All enquiries to Caradon-Ideal Ltd
Strebel Ltd	1F Albany Park Industrial Estate, Frimley Road, Camberley, Surrey, GU15 2PL
	Tel: 01276 685422 Fax: 01276 685405
Sugg	All enquiries to Potterton Myson Gateshead Office
Sunvic Controls Ltd	Bellshill Road, Uddingston, Glasgow, G71 6NP
	Tel: 01698 812944 Fax: 01698 813637
Superswitch Elec. Ltd	Holdsworth Street, Redditch, Stockport, Cheshire, SK5 6BZ
	Tel: 0161 431 4885 Fax: 0161 431 4385
Switchmaster Controls	See Appliance Components Ltd
Teddington Controls Ltd	Holmbush, St Austell, Cornwall, PL25 3HS
	Tel: 01726 74400 Fax: 01276 67953
Thermecon Boilers Ltd	Melton Road, Melton, Woodbridge, Suffolk, IP12 1NH
	Tel: 01394 386699 Fax: 01394 386609
Thorn Heating	All enquiries to Potterton Myson (Gateshead Office)
Thorn Security Ltd	206 Bedford Avenue, Slough Industrial Estate, Slough, SL1 4RY
	Tel: 01753 571719 Fax: 01753 571721
T.I. Glow Worm	All enquiries to Hepworth Heating
T.I. Radiation	All enquiries to Hepworth Heating
Time & Temperature	Unit 56, Plume Street Industrial Estate, Plume Street, Aston, Birmingham, B6 7RT
	Tel: 0121 327 2717 Fax: 0121 327 2395
Timeguard	Waterloo Road, London, NW2 7UR
	Tel: 0181 450 8944 Fax: 0181 452 5143

Manufacturers' trade names and directory

Tower	See Grasslin (UK) Ltd
Trac Time Controls	Thirsk Industrial Park, Thirsk, North Yorks, YO7 3BX
	Tel: 01845 526006 Fax: 01845 526010
Trianco Redfyre Ltd	Thorncliffe, Chapeltown, Sheffield, S30 4PZ
	Tel: 0114 257 2300 Fax: 0114 257 1419
Trisave	All enquiries to Crosslee PLC
UGI Smith Meters Ltd	Heating controls, see Appliance Components Ltd
Unidare	See Seagoe Technology
Vaillant Ltd	Vaillant House, Trident Close, Rochester, Kent, ME2 4EZ
	Tel: 01634 290155 (service) 0870 6060 777 Fax: 01634 290166
	(technical) 01634 292392
Venner	All enquiries to Heating Control Services
	Tel: 01922 34503
Viessmann Ltd	Hortonwood 32, Telford, Shropshire, TF1 4EU
	Tel: 01952 670261 Fax: 01952 670103
Vokera	See G R Claudio Ltd
Vulcan	All enquiries to Caradon-Ideal
Warmworld UK	See Prism Energy Technology Ltd
Wickes (boilers)	All enquiries to Halstead Heating
Wickes (controls)	All enquiries to Landis Staefa
Wilo Salmson Pumps	Ashlyn Road, West Meadows Industrial Estate, Derby, DE2 6XE
	Tel: 01332 385181 Fax: 01332 344423
Henry Wilson	All enquiries to Potterton Myson (Gateshead Office)
Worcester Heat Systems Ltd	Cotswold Way, Warndon, Worcester, WR4 9SW
	Tel: 01905 754624 (technical service) 0990 266241
	Fax: 01905 754619 (technical service) 01905 757224
Yorkpark Ltd	16 St George's Industrial Estate, White Lion Road, Amersham, Berks, HP7 9JQ
	Tel: 01494 764031 Fax: 01494 765754

Index

ACL
   See also ACL-Drayton
   clockwatcher, 164
   cylinder thermostats, 74
   motorised valves, 91–92, 202
   programmers and time switches, 4–7, 215, 218, 219
   programmers and time switches with inbuilt or external sensors or thermostats, 62–64
   room thermostats, 78–79
ACL-Drayton
   programmable room thermostats, 64–65
   room thermostats, 79
Appliance Components Ltd
   See ACL
Ariston
   boilers gas, 109
AWB
   boilers gas, 109

Barlo
   boilers gas, 110
   cylinder thermostats, 74
   motorised valves, 92
   programmers, 7, 215
   room thermostats, 79
Baxi
   boilers gas, 100–112
Benefit
   cylinder thermostats, 74
   room thermostats, 79
BKL
   boilers gas, 112
Boulter
   boilers gas, 112–113
   boilers oil, 156–157
Brassware Ferroli
   room thermostats, 79
   boilers gas, 116–117

Centurion
   boilers gas, 113

Chaffoteaux
   boilers gas, 113–114
Combi Company
   boilers gas, 114
Crossling
   programmers, 7, 216

Danfoss
   See also Danfoss-Randall
   BEM 4000, 165
   BEM 5000, 166
   burner control boxes, 162
   control plan 1.1, 190
   control plan 2.2, 203
   control plan 2A, 190
   control plan 2C, 203
   cylinder thermostats, 75
   motorised valves, 92–93, 203
   programmers and time switches, 7–8, 217, 218
   room thermostats, 79–81
Danfoss-Randall
   See also Danfoss
   See also Randall
   programmers and time switches, 8–9, 215, 216, 218, 219
   room thermostats, 79
   WP75H programmable hot water thermostat, 65
Dataterm
   optimiser, 168
Drayton
   control plan 1, 203
   control plan 2, 184
   control plan 5, 204
   control plan 7, 205
   cylinder thermostats, 75
   digistat prgrammable room thermostats, 65
   motorised valves, 93–95, 203–205
   programmers and time switches, 9–10, 216, 219
   room thermostats, 81
   SU1 switch unit, 190

Index

    SU2 switch unit, 191
    theta autotherm, 169
DS
    burner control box, 62

Eberle
    cylinder thermostats, 75
    programmers and time switches, 10–11, 58
    room thermostats, 81
    RTR clock thermostat, 65–66
Ecko
    room thermostats, 81–82
Elestra
    burner control box, 162
ELM Leblanc
    boilers gas, 115
Eurocombi
    boilers gas, 115

Firefly
    boilers gas, 117
Flash
    programmers and time switches, 11

Gemini
    boilers gas, 117
    boilers oil, 157
Geminox
    boilers gas, 117
Gledhill
    boilermate, 170
    cormorant, 170
    gulfstream, 171
Glotec
    boilers gas, 118
Glow-worm
    programmers, 12, 216
    boilers gas, 118–122
Grant
    boilers oil, 157
Grasslin
    programmers and time switches *see* Tower

Halstead
    boilers gas, 122–124
Harp
    programmers, 12
Hawk
    time switch, 12
Heatwave
    programmers, 12

Homewarm
    manual system, 205
    auto system, 206
Honeywell
    AQ 6000, 171
    burner control box, 162
    CM programmable thermostats, 66
    control plan A, 185
    control plan C, 187
    control plan G, 197
    control plan L, 187
    control plan S, 195
    control plan Y, 195, 206–207
    cylinder thermostats, 75
    motorised valves, 95-96, 197, 206–207
    programmers and time switches, 12–14, 216–217, 219
    room thermostats, 82–84
    Y604A Sundial plan, 174
    Y605B Sundial plan, 175
Horstmann
    centaurstat programmable thermostat, 67
    cylinder thermostat, 75
    motorised valves, 96
    programmers and time switches, 14–24, 58, 61, 108, 216–218
    room thermostats, 84
HRM
    boilers oil, 157–158
Perrymatic
    boilers oil, 158

Ideal
    programmer, 24
    boilers gas, 124–128
International Janitor
    programmer, 24

KDG
    room thermostats, 84
Keston
    boilers gas, 128

Landis & Gyr
    See also Landis & Staefa
    burner control boxes, 162
    chronogyr, 67–68
    cylinder thermostats, 75
    LGM system, 297
    motorised valves, 97, 207
    programmers and time switches, 24–27, 216, 218
    room thermostats, 84

Index

Landis & Staefa
   chronogyr, 68–69
   programmers and time switches, 25, 216, 218

Malvern
   boilers gas, 128
Maxol
   boilers gas, 128, 129
Myson
   See also Potterton
   boilers gas, 129–131
   motorised valves, 97–98
   programmers, 27

Nettle
   room thermostats, 84
Nu-Way
   boiler control boxes, 162

Ocean
   boilers gas, 132

Pegler
   Belmont tec, 177
   room thermostat, 84
Petercem
   burner control box, 162
Potterton
   boilers gas, 132–136
   cylinder thermostats, 76
   motorised valves, 98
   PET programmable thermostats, 69
   programmers and time switches, 27–31, 216–219
   room thermostats, 84–85
Powermatic
   boilers gas, 136–137
Powermax
   boilers gas, 137
Pro-Heat
   programmers and time switches, 31
Proscon
   cylinder thermostat, 76
   room thermostats, 85

Radiant
   boilers gas, 137
Randall
   See also Danfoss Randall
   cylinder thermostats, 76
   EBM 2.1, 178
   motorised Valves, 98
   programmers and time switches, 31–38, 58–59, 61, 207, 216–218
   room thermostats, 85–86
   TP programmable thermostats, 69–70
Ravenheat
   time switch, 38
   boilers gas, 138
Rayburn
   boilers gas, 139
Ringdale
   801, 180
   802, 180

Sangamo
   programmers and time switches, 38–44, 59, 61, 207, 218
   room thermostats, 86
Satchwell
   See also Sunvic
   burner control box, 162
Saunier Duval
   boilers gas, 139–142
Sauter
   room thermostats, 86
Selectos
   burner control box, 162
Sime
   boilers gas, 142
Sinclair
   boilers gas, 142
SMC
   control pack system, 201, 208
   programmers, 44, 61
Smiths
   centroller, 70–71
   cylinder thermostats, 76
   ERS programmable thermostat, 71
   motorised valves, 99
   programmers and time switches, 44–47, 60, 216
   room thermostats, 86

SOPAC
   cylinder thermostats, 76
   motorised valves, 100
   programmers, 47
   room thermostats, 86–87
Southern Digital
   time switch, 47
Stewart
   burner control box, 162
SUGG
   programmer, 47

Index

Sunfine
   room thermostat, 87
Sunvic
   clockbox, 181
   control system 4, 195
   cylinder thermostats, 76–77
   duoflow systems, 209, 210
   EC clock thermostat, 71
   motorised valves, 100–104, 191, 209–210
   programmers and time swtiches, 48–50, 216, 219
   room thermostats, 87–88
   TLC clock thermostat, 72
   unishare system, 195
Superswitch
   programmers and time switches, 50
Switchmaster
   cylinder thermostat, 77
   midi system, 211
   motorised valves, 104, 105, 206, 211
   programmers and time switches, 50–52, 60, 217
   room thermostats, 88–89
   serenade programmable thermostat, 72
   symphony, 72

Teddington
   burner control box, 162
   cylinder thermostat, 77
   room thermostat, 89
Thermoflex
   burner control box, 162
Thermomatic
   boilers gas, 142
Thorn
   boilers gas, *see* Myson
   boilers oil, 158
   programmers, 52, 217
Thorn Security
   room thermostat, 89
Tower
   cylinder thermostat, 77
   motorised valves, 105
   programmers and time switches, 52–54, 216, 219
   room thermostat, 89

Trac
   cylinder thermostats, 54
   programmers and time switches, 54
   room thermostat, 89
Trianco
   boilers gas, 142–143
   boilers oil, 158–160
   burner control box, 162
Trisave
   boilers gas, 143
Tristat
   room thermostat, 89

Unity
   room thermostat, 89

Vaillant
   boilers gas, 144–148
   room thermostats, 89
   VRT calotrol, 72–73
Venner
   programmers and time switches, 55–57, 60
Vokera
   boilers gas, 148–149
   room thermostat, 90

Warm World
   boilers gas, 149
Wickes
   boilers gas, 150
   cylinder thermostat, 77
   motorised valves, 105
   room thermostat, 90
Worcester
   boilers gas, 151–154
   boilers oil, 160–162
   room thermostat, 90
Wylex
   room thermostat, 90

Yorkpark
   boilers gas, 154–155